Once Upon a Flock

Flock

Life with My Soulful Chickens

Lauren Scheuer

Souvenir Press

For Sarah

Warwickshire County Council

NOV 9/14			
27/7/15			
22/8/15			
11 OCT 2018			

This item is to be returned or renewed before the latest date above. It may be borrowed for a further period if not in demand. **To renew your books:**

- Phone the 24/7 Renewal Line 01926 499273 or
- Visit www.warwickshire.gov.uk/libraries

Discover • Imagine • Learn • *with libraries*

Warwickshire
County Council

Working for
Warwickshire

014007226 5

Contents

Once Upon a Flock

Backyard Makeover

With a yard like ours, there is little reason ever to be indoors. Our home in Massachusetts is surrounded by forest, with classic New England granite ledges and boulders and wetlands. When our daughter, Sarah, was small, we hunted blueberries.

We dissected owl pellets, caught garter snakes, and built leafy homes for pet caterpillars. We scooped tadpoles and tiny frogs out of the vernal pool in the woods and raised them in an aquarium on our back stoop.

We adopted Marky the terrier, an independent fellow who also enjoys outdoor adventures in all seasons.

While Marky patrolled the yard and the forest, Sarah and I continued to explore and create. We made forts out of sticks, and we hung rope swings from the trees. As my girl grew, so did our building projects. I built a tree platform for Sarah and her friends, and I was thrilled to find a new use for my beloved collection of power tools.

Back when Danny and I were newlyweds in Boston and were graduating from milk crates and trash-day sidewalk finds to respectable grown-up furnishings, I had made many a bookshelf and table with these old tools.

For a while, when I had no idea what to do with my newly acquired fine art degree, I considered furniture making as a career.

Then I took some woodworking classes in which I learned that I should not consider furniture making as a career. My attention span is too short. I can never find a ruler or a level when I need one, so my final product ends up looking a lot like the five-second sketch I did just before starting to build whatever it was I just sketched. The excitement for me is in making a sketch of a chair and then turning it into a real live chair that I can sit on. To sketch a bookcase, and then—voilà!

A real bookcase is born. And really, who needs a ruler and a level when you've got a few hundred drywall screws and a big tub of wood putty? I possess the talent to skillfully disguise my mistakes, no matter how huge.

The tree platform I built for Sarah and her friends was so liberating that way.

Fine details were unnecessary. The kids didn't stand back and critique my work; they just climbed all over it. They wanted pulleys and ropes and ladders and more pulleys. So that's what I gave them, along with trapdoors and secret compartments.

It was a solid structure, albeit a little precarious, perched eight or nine feet above a rocky outcropping. So as a precaution I threatened to punish anybody who got hurt.

I stretched a zip line from one end of the yard to the other, again with the threat of severe punishment for any broken bones or concussions. As a result, we had no serious injuries or deaths, and for many years this backyard was a mecca of colorful adventures and activity.

Then one day, totally without warning, Sarah grew up. She retreated to her room and began morphing into a teenager—the sedentary, electronic type.

Carpenter ants and paper wasps took up residence in the tree platform and went to work reducing it to mulch. A forest of tiny pine trees sprouted in the sandbox. The zip line rusted and sagged. Soccer balls and hula hoops disappeared under autumn leaves.

I stood alone, slump shouldered, in my echoing wasteland of a yard. Well, I wasn't totally alone. My faithful canine whacked the back of my leg with his squeaky toy.

Marky was always ready for fun, but at that moment I wasn't. And besides, Marky only wanted to play his own game, with his own rules. And he always won.

Eventually he dropped his toy and chased a chipmunk into the old woodpile. Something had to change. My yard needed a serious jump start.

I thought about my friend Patricia, who lives down the road and whose yard is full of life. Whenever I visit Patricia, she and I wander among her perennials. We sip tea and discuss the year's tomato crop. In her yard, children frolic, sheep graze, and colorful chickens drift in and out of the garden. In Patricia's yard, the sun shines brighter and the grass is always greener.

I didn't necessarily want to have a farm. I've never pined for a flock of sheep, nor have I ever felt great desire for a herd of children.

Chickens, on the other hand . . . Chickens might be just what I needed. Chickens would bring my yard back to life. I would get my own flock of wonderful birds, and my family would come skipping out into the sunlight to enjoy them with me. And even if my family never emerged, I would have a bunch of birds to call my own. The thought delighted me.

And fresh eggs! My birds would give me eggs—what a bonus! Backyard eggs would be my ticket out of the factory farm conundrum. No longer would I have to buy supermarket eggs and feel guilty about it, knowing that the chickens who produced them had most likely never seen sunshine or enjoyed fresh grass or bugs.

If all these reasons weren't enough, my flock of chickens would necessitate a new building project. I would design and build the perfect coop.

It didn't take much chatting with Patricia to get me hooked on the idea. She assured me that raising chickens was both easy and fun. As a matter of fact, Patricia was thinking of getting some new chicks in the spring too. So we planned our order together and discussed a schedule.

If we ordered our chicks to arrive in February, they would grow up and lay eggs by the end of the summer. That sounded good to me. February was three months away, plenty of time to study and plan and learn all I could.

I began with the coop.

2

The Coop

Chicken coop. A simple concept.

Once I had scanned a few library books and chicken websites, I pretty much had the gist. The coop and henhouse keep the chickens safe from predators. In the henhouse, chickens roost by night and lay eggs by day. In the coop, they scratch and peck and take in the fresh air and sunshine and wait impatiently for me to come and let them out. I tore open a fresh ream of paper, picked up a pencil, and started drawing.

Sketches flew.

I pulled ideas from exotic tales,

faraway lands,

art genres,

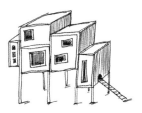

Greek drama and ancient empires,

and my favorite foods.

Ideas came faster than I could scribble them down.

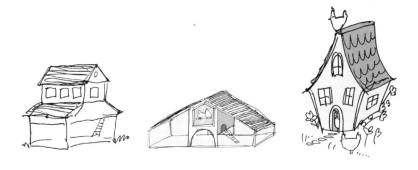

Sketches stacked up on the kitchen table and blew across
the kitchen floor.

I stirred pasta with one hand and drew coops with the other.

Danny and Sarah stepped over piles of coop books and shuffled across mounds of drawings, and wisely kept on walking. They were familiar with my tendency to obsess and watched from a safe distance.

I hadn't really discussed the chicken plan with my family. We're not the family-meeting type. But since so much evidence had accumulated, it was certainly not a secret, and I felt that if they had any concerns, they could attempt to discuss them with me. And if they chose never to chat about it with me, then hey . . . No conflict? No problem.

In my free time I wandered through lumberyards, breathing deeply the smell of fresh sawdust and examining lengths and grades of pine and plywood. I visited the local hardware store to fondle latches, bolts, and hinges. I hunted down the most economical materials, and I planned my coop's dimensions based on the length of wood that could squeeze into my Honda Fit: eight feet max from hatchback to dashboard. With these restrictions, I came up with just the right design.

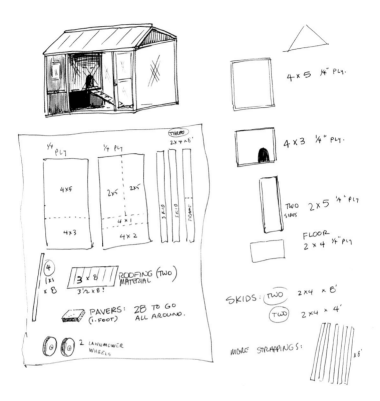

Attractive, sturdy, and with wheels on one end so I could move it around the yard. Secure, easy to clean, and with a doorway just wide enough to fit my wheelbarrow.

Perfect for a small flock.

3

Chickens 101

February crept up, and suddenly it was time to place the chick order. I swept aside mountains of coop sketches in order to barrel into this chicken thing.

It was more complicated than I had expected.

There were so many breeds to choose from, and so many considerations. Fancy birds are not necessarily the best egg layers. Silkies and Sebrights and bantams can be tricky to raise and care for in a mixed flock. And those hens with the big poufs on their heads? I read that they can't see a darn thing—they tend to run around bumping into walls.

Really, I just wanted a few colorful egg-laying chickens. Nothing fancy. But even among basic chickens, there are dozens and dozens of breeds to choose from. These chickens were going to be my living lawn ornaments, so color was the first consideration. They had to complement the peony and the coneflower, the sedum and the phlox. I wanted a variety of birds, I wanted good layers, and I needed hens that wouldn't freeze to death in a New England winter.

I settled on a black, a striped, a yellow, and a red one.

Rhode Island Red and Barred Plymouth Rock originated right here in the Northeast. They would know how to handle the harsh winters.

My Black Australorp would glimmer in the sunshine like an iridescent beetle.

And the Buff Orpington? Well, I just wanted a reason to say "Buff Orpington" every day. *Buff Orpington, Buff Orpington, Buff Orpington.*

I wanted only three chickens, because three chickens constitute a flock, and three hens would probably provide enough eggs for our family of three people. But I figured I ought to order the fourth one for good measure, because raising chickens has its risks.

Every night I read about these risks in my chicken books, and every day I worried about everything I had read the night before.

My chicks could drown in their water bowl. They could die of heat stroke; they could die if they got too cold. They could sleep in a pile and suffocate each other. They could get pasty butt when a nugget of dried poop gets stuck like a cork and you have to dislodge it with a wet towel or a bit of olive oil, or else they will die from, I guess, internal combustion. And if the floor is too slippery, their feet could splay, and I would have to make miniature leg splints out of toothpicks and tape. And if they survived this gauntlet of horrors in their little chickie lives, they could still drop dead at any time for no apparent reason.

It's true. I read it on the Internet.

If I managed to keep them alive all the way to adulthood, more dangers lurked in their future: they could get sour crop, bumblefoot, fowl pox, and egg drop.

BUMBLE

They could get eaten by foxes, skunks, hawks, owls, raccoons, weasels, fishers, or neighborhood dogs. They could go broody, which is when a hen decides she wants to be a mother, or they could come down with any number of fatal ailments. And worst of all, they could grow up to be roosters.

Patricia and I planned to pay extra for our chicks to be "sexed" at the hatchery by a professional chick checker who can tell a girl from a boy, but this is not an exact science. Mistakes are known to happen. So the fourth chick in the order was my insurance. Although at four dollars apiece, I wasn't risking much as far as money was concerned.

Just like with the coop plans, I didn't involve Danny and Sarah directly in my chicken plans. They could see what I was doing. Every night we scooted stacks of chicken books to the far side of the kitchen table in order to eat dinner, and every day those books were scattered all over the table again. Lists and sticky notes hung out of every book and were taped all over our walls as well. If Danny or Sarah had any concerns about my plan or wanted to know the details, they could have asked me.

They did occasionally mutter some comments, opinions, and feeble arguments against chickens.

I listened politely and then flipped open another chicken book.

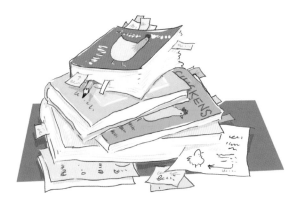

When I officially informed Danny that chickens were inevitable, he rubbed his forehead, let out a faint groan, and initiated one mature and rational discussion. He said he didn't see the need for more pets. We already had a dog, two guinea pigs, and the Betta fish that Sarah had won at a birthday party. He felt that our lives were already plenty complicated, and didn't I often complain of not having enough time to dedicate to my freelance illustration business?

I responded with silence.

"Why do you want chickens?" Danny asked.

"Because I want the experience."

Long pause.

"Okay," he said.

I called Patricia and we placed our order that night.

4

Fleeting Cuteness

Our bundles of fluff arrived by mail. Day-old chicks can be shipped successfully because they are born with enough nutritional reserves to last three days. In a real broody nest situation, this makes it possible for all the chicks to hatch over the course of a day or two without the mother hen needing to leave the nest to feed her early hatchers.

With the current backyard chicken craze under way, postal clerks are familiar with peeping packages. They call recipients immediately to pick up their parcel, and chicks arrive at their new homes a little shaken but in good physical condition.

Patricia picked up our chicks at the post office, and I met her at her house to sort them out. The Barred Rock and the Australorp chicks had distinctive coloring, so we knew for certain which ones they were. Most of the other breeds were tough to distinguish—they were all variations on the basic yellow puffball—so we made our best guesses, and I took my four babies home.

I arranged them on a bed of paper towels and straw in a plastic bin in the living room. Danny and Sarah stumbled in to take a look and were instantly smitten. All three of us were. At two days old, these little birds tottered nonstop, bumped into each other, sat down, stood up, and jettisoned about . . . and stole our hearts.

Our three giant human heads hovered over the chicks' bin, huge smiles on our faces. Within minutes, Sarah had given them names. That evening, we ate a picnic dinner in front of the chick bin, and that night Sarah serenaded them with soft guitar chords.

The chicks stood quietly in the warm glow of their heat lamp until they tipped over, one by one, right where they stood.

The next morning after Sarah had left for school and Danny for work, I called my illustration client to tell her that something had come up—I needed to extend the deadline another day. Then I poured my chicks out onto the carpet and wallowed in them.

I discovered right away that each one came equipped with a unique personality.

Little White was docile and just plain sweet. She stayed willingly wherever she was placed: on my finger, on my shoulder, upside down, right side up. Sarah had named her Little White because the first tiny feathers that had come in were white wing feathers. We thought we'd gotten a chicken mixup and that she would turn into some sort of white-feathered chicken that we hadn't ordered. When we later realized that she was, in fact, a Buff Orpington, the name Little White still seemed right for her sweet demeanor, so she kept her name.

In fact, because of her calm and respectable comportment and her blossoming good looks, her name morphed into Lil'White, spoken with a soft Southern accent. She was a belle if ever there was one. And she became Sarah's favorite.

Lucy, the Barred Plymouth Rock, seemed curious and thoughtful and a bit anxious. But as long as we didn't make any sudden moves, she was happy to hang out on my hand. Every day I watched with anticipation for her striking barred plumage to emerge.

And as her feathers came in, she began to look like a baby robin.

Hatsy started out as an elegant golden chick.

But she began to grow very quickly and took on a hawklike appearance that was a bit disconcerting. We worried from the start that she might be a he. Regardless, we realized she was neither the Rhode Island Red nor the Buff Orpington that we had expected her to be. In fact, we couldn't figure out what breed she was. But we were fine with that mystery and just kept our fingers crossed that she would grow up to be a hen and not a roo.

Hatsy stepped into a leadership role right away. She was smart and quick and adventuresome, and she set off on solitary reconnaissance missions across the living room whenever the opportunity arose.

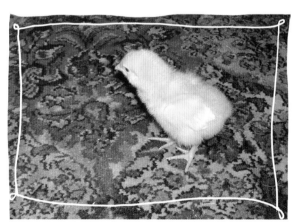

The fourth chick, Jenny, was our Black Australorp. Her coloration was striking, and her personality was . . . well . . . Jenny was a screaming, wailing, inconsolable baby.

When separated from the other chicks, even for a moment, she fell completely apart. Jenny was as easy to love as she was to laugh at.

As my obsessions rolled predictably along, I fell headfirst into my new flock. I picked up my camera

and had a difficult time

putting it down.

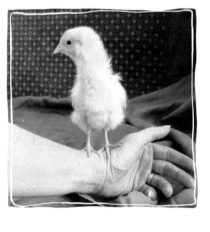

I spent every minute watching and playing with the babies, and I clicked my camera in rapid-fire mode.

Clearly our chicks were skilled in mind control. We melted under their powers. Dinner boiled over, homework went undone. Some afternoons I forgot to pick up Sarah from school. "I was running errands," I explained, plucking bits of chick bedding out of my hair.

Marky watched our chicks with quivering lips.

I tried to introduce them to him as the books had advised, but this just made him drool and whimper. I had to face the fact that he was a terrier, bred to hunt, and that all things small and fuzzy are prey to him. As well trained as Marky was, he didn't have the ability to control himself around these chicks. So whenever we took the chicks out to play on the floor, Marky was banished to the backyard.

The chicks quickly outgrew their bin. I scrounged in the attic among Danny's lifetime collection of boxes and cartons and chose the large Sony Trinitron box that had waited nearly a decade for its purpose to come along.

The box gave them plenty of space to run around and spill their water and poop in their food, which they did nonstop. I felt it was time for the chicks to learn a thing or two about being chickens, so I built them a training roost.

They caught on to the roosting thing right away.

I pasted pictures of grown chickens on the wall in front of them,

so they could marvel at what they would become.
I cut a window and fitted it with clear plastic

so we could watch the chicks,

and they could watch us.

When they outgrew their Trinitron box, I cut a doorway and duct-taped another room to it. I added another table and another lamp to the setup, and that was pretty much the end of our living room's feng shui.

I kept snapping photos, hundreds a day, desperate to preserve the last ounce of cuteness for our memories.

And then all of a sudden, they turned hideous.
Well, all except Lil'White.

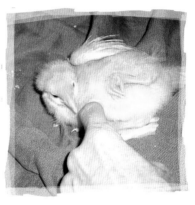

She remained cute as a button.

Cuteness has its price, however.

Lil'White had to put up with extra love and attention from Sarah and her friends because of her adorability factor.

Meanwhile Hatsy, Lucy, and Jenny had entered their grue-

some stage. Scruffy feathers bulged out in all directions.

Red and white plumage emerged from Hatsy's yellow fluff, and we still couldn't figure out what breed she was. Sarah thought she might be a dinosaur; I was pretty sure she was a rooster.

The chicks' gangly reptilian toes grew wicked toenails. Their poops, once cute nuggets, were now atrocious, reeking turds.

Birds form new feathers within a sheath, like a tiny drinking straw, and as the girls' feathers grew, the sheaths crumbled into flakes and dander and dust.

The girls flapped around and ran laps inside their boxes, sending clouds of dust billowing all over the living room. And the dust stuck. To the walls. The furniture. The curtains. The ceiling. Us.

Friends still stopped by to see our darling chicks, but now they only stood in the doorway. They eyed the girls from across the room and then remembered that they couldn't stay long.

Our house smelled like a barn. It was time to build that coop.

5

Henhouse

I descended to our basement garage and fired up my beloved power tools. I worked for three days and three nights, enjoying every whir of the drill, every whiff of sawdust, every Sponge-Bob bandage applied to my bloody wounds.

My design on paper was a rough one, and I hadn't solved all the construction issues, but I played it by ear. I built the two long sides first, then the two short sides.

I had read that chicken wire is not a deterrent to some predators. Snakes and small rodents can squeeze through it, and raccoons can rip right in and snatch a panicking bird. So, instead, I chose to use hardware cloth, a welded wire mesh that is utterly impenetrable. It's also about as dangerous and unwieldy to work with as razor wire.

I rolled out the hardware cloth to measure and cut it, and it curled right back onto itself and me, over and over again, piercing through all layers of clothing and even through my shoes. It took forever to cut the stuff and to fasten it to the coop, and it took just as long to stop the flow of blood from my countless gashes, scrapes, and puncture wounds.

But once it was in place, the mesh looked tidy and attractive, and I felt confident that this would prevent even the most crazed and ravenous predators from harming my girls.

When the four walls were finished, I dragged them out of the garage into the April sunshine to prop them up and put them all together.

My hastily sketched plans, however, hadn't addressed this phase of construction. I had no idea how to put these walls together. I went back to the wondrous hardware store, took another fanciful jaunt up and down the aisles, and discovered exactly what I needed: metal brackets. Why had nobody ever mentioned these in woodworking class? Precise joinery and fine dovetails have their place, but this coop, my biggest building project ever, was going to stand erect only by the grace of hundreds of drywall screws and these awesome metal brackets.

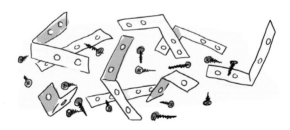

I bought twice as many as I needed and was home in no time slapping those walls together. Moths danced around my head under the driveway spotlight, and Sarah and Danny watched me from the kitchen window, smiling and waving their approval with plastic spoons as they ate their dinner of cold cereal. I came inside at bedtime, then went back out the next morning and assembled the roof, making good use of those extra metal brackets. I planed and sanded the roosts so my girls wouldn't get splinters in their delicate toes. I gave their ladder plenty of rungs for easy climbing.

In the garage, I opened all the cans of leftover house paint, lined them up, mixed this one with that one, and added a splash of Sarah's acrylic craft paint to achieve the right hue. All the painting was done by nightfall. Since the coop needed to dry overnight, our family got to enjoy one last sentimental evening with the chickens in the living room. Hatsy stood on my shoulder and picked at the dried paint in my hair. Lucy gave herself a good shake, emitting a final massive explosion of dust. Jenny and Lil'White wandered together into the den and pooped on the floor.

The next morning I wheeled my spectacular mobile coop onto the lawn and brought the girls out to experience sunshine and a cool breeze for the first time.

Inside their beautiful coop, the girls cowered under a plastic stool.

Outside the coop, Marky stared at them and drooled.

I went in and sat down on the stool to help the girls adjust to their new environment. They quickly loosened up and began to explore.

Lil'White figured out how to tip over the feeder.

Hatsy gave my eyebrows a nice plucking.

With Marky's encouragement, Hatsy learned to run up the ladder.

Getting back down was more of a challenge.

Lil'White couldn't figure out the ladder at all, so I lifted her to the top and we did a little practicing.

That night, the girls slept in their new home, and I lay wide awake in my bed. Was the coop secure enough? Were the girls too cold? Would the heat lamp catch everything on fire? Had I latched the door? I got up and tiptoed out into the night to check on them. Twice.

And I was twice relieved and twice proud to find that all was well and that the flock was safe and comfortable. I had built them a nice home.

The next day I wheeled the coop to a new spot so the girls could have fresh grass, and the wheels fell off. But the rest of the coop was sound, so Danny helped me drag it to its permanent location at the far end of the yard.

I was only momentarily disappointed at the failure of those wheels, for now that the coop was immobile, a new mobile coop was in order. A new building project!

A "chicken tractor"—a lightweight structure on wheels— is what the girls needed. I had seen photos of these tractors online and had already done some sketches. I even had enough leftover lumber and metal brackets to make one.

And so I did. It only took an afternoon.

In their new chicken tractor, the girls could scratch and peck wherever I wanted them to—and without getting their heads stuck between Marky's teeth.

The flock's very first summer was approaching, and new excitement was popping up every day. The girls discovered tangy dandelions, sweet clover, and lemony wood sorrel growing thick among the few blades of grass that I call my lawn.

Hatsy, always on the move, chased mosquitoes and moths

and plucked them right out of the air. Lucy learned that worms taste like candy and that yellow jackets do not. Lil'White had no interest at all in worms. She preferred the satisfying crunch of a fresh beetle.

The girls were intrigued with my rakes and shovels and gardening gear. It was just a thrill a minute for these young ladies.

And the girls *were* young ladies. No longer chicks, they were now what you call pullets: young hens. They sported beautiful mature feathers, and their combs and wattles were starting to show. And while we had feared from the start that Hatsy might be a rooster, it was Jenny who crept up and surprised us.

Yes, Jenny, the screamingest cryingest baby.

6

Summertime

We had tried to ignore Jenny's good-sized comb and wattles, but her feet were huge too, and her tail feathers were getting longer. Jenny outgrew all the crying and carrying on and adopted a proud and protective role. She wasn't crowing yet, but we could no longer deny that she was a he.

Our town has very few written restrictions when it comes to keeping farm animals, even roosters. Farming is a part of Upton's historic pride, and people here seem to take pride in our town's lack of rules as well. Patricia had her chickens. She also had a llama, a flock of sheep, and several horses. Other friends across town kept rabbits and goats. Our neighbors had no problem at all with our raising a small backyard flock—in fact they were looking forward to tasting our homegrown eggs. But I just could not justify keeping a noisy rooster. I liked the neighbors too much to risk having them all hate me.

Jenny had to go.

Sarah was adamant that we find our rooster a proper, happy home. I agreed and assured her that there was no way I could chop Jenny's pretty little head off, or allow anyone else to do such a thing. We would find Jenny a farm where she could live happily ever after.

I began searching websites like craigslist for someone who was looking for a nice rooster. Unfortunately, Jenny had a lot of competition in the rooster re-homing market. Craigslist displayed listings for dozens and dozens of roosters "free to a good home." Trying not to feel discouraged, I continued my hunt. And as Jenny's prospects for safe haven dwindled, her comb and wattles just kept on growing.

Soon she would begin to crow. Loudly. Day and night. I was sure that our once amiable neighbors would band together, a sleep-deprived mob brandishing knives and forks and bottles of chicken marinade, and I would barricade myself inside the coop with Jenny and the flock for our final standoff.

I continued to lose sleep over Jenny's fate and our own

until Sarah and her class went on a field trip to Farm School. Farm School is a working farm in Athol, Massachusetts, where groups of schoolchildren spend several days wallowing in farm life, right up to the very rims of their muck boots.

Sarah and her friends tended to the chickens and the pigs. They gardened. They drove oxen. They kissed calves, they planted seeds, and they shoveled all sorts of poop. Each night the kids fell exhausted into their bunks, and each morning they rose with the sun and did it all again.

It was a good old-fashioned idyllic learning experience.

When their time at Farm School was over, I drove out to pick up Sarah. I toured the farm and was impressed with the lovely flocks of chickens and their accommodations. I chatted with various staff members and casually asked if they might have room at the farm for a handsome Australorp rooster named Jenny. One young fella, Nick, told me he was starting a flock of his own, and he thought a Black Australorp roo might make a nice addition.

What a wonderful coincidence! I just happened to have such a rooster sitting in the backseat of my car. I opened the door, Jenny hopped out, Sarah and I jumped in, and we tore out of there before Nick could change his mind. Sarah and I wept a few tears on the way home. Sarah also pitched a fit, furious with me for giving up our beloved Jenny so abruptly. But deep down she knew that I had done it in Jenny's best interest, and we were both comforted by the thought of proud Jenny heading up her own—I mean *his* own—herd of hens.

Back at home, Hatsy, Lucy, and Lil'White seemed to miss Jenny for a few days, but they rearranged themselves into a tidy flock of three and created a comfortable sisterly order. A pecking order.

Hatsy, the leader from the start, retained her top position. Lucy assumed position number two, and Lil'White settled for the spot at the bottom. I was pleased to find that their creation of a pecking order involved no pecking. The girls were very kind to one another, although Hatsy was a bossy gal and demanded dibs on the finest treats.

Hatsy and Lucy seemed pretty chummy, while Lil' White remained aloof. But all three enjoyed doing chicken things together as a good-natured flock.

I had assumed that my flock of chickens would spend carefree days just milling about, each one oblivious to the others. They'd peck and scratch and bump into each other and move on. So I was intrigued when I noticed that Lucy and Hatsy seemed to enjoy each other's company.

These two girls sought each other out. They strolled around together, chatting agreeably about this and that. Lucy would reach over to straighten a feather on Hatsy's back. Hatsy would pause and exclaim appreciatively. Then Hatsy would peck a crumb of food from the side of Lucy's beak like a lady tenderly wiping a gob of potato salad off her friend's cheek.

Lil' White mingled somewhat too, but she was an independent gal. She seemed comfortable among her friends but spent more time with her back to them than otherwise. Her Orpington plumage was breathtaking. I think Buff Orpingtons might just have twice as many feathers as other chickens.

And she never had a feather out of place, never a speck of poo on her voluptuous behind. Scratching and pecking all day and traipsing through goodness knows what with her bare toes, Lil'White's feet remained pearly and spotless. She was the first one our friends noticed when they came to see my flock.

And she remained Sarah's very favorite chicken.

Sarah couldn't help but pick her up and smile into those vapid Orpington eyes.

Hatsy, who began her life as the most enormous chick, grew up to be the smallest hen in the flock. A dizzying sight, always in motion, Hatsy was the hunter.

When she had cleaned up all the bugs on the surface of the lawn, she started hunting beneath. With lightning-fast scratching of her pointy little toenails, she overturned great clumps of sod. Everywhere I wheeled the chicken tractor, Hatsy initiated a new excavation project, and she gouged ankle-twisting divots all over the yard.

Bugs weren't her only quarry. I tried to remain calm the day Hatsy tore across the lawn with a snake squirming around her head.

Lil'White gave pursuit, but I don't know what happened next because I couldn't look.

My little orange dynamo Hatsy also laid our very first egg. She was way ahead of schedule. We had expected the girls to begin laying in their fifth or sixth month. Hatsy's first egg came in her fourth month, completely unexpected.

As soon as she emerged from the nest box that memorable day, I peeked inside to see what on earth she had been up to in there.

I gasped when I saw it. I picked it up as if I had discovered the rarest of treasures. It was breathtakingly beautiful. It was egg shaped. It had an egglike color. It felt like an egg. It had some nice random speckles on it. It was . . . *warm*.

This first egg was far more exciting than I had ever expected it to be. The immense thrill that overcame me was embarrassing, even to myself, standing alone beside the nest box. I thanked Hatsy with sincerity, and cradling the egg in both hands, I carried it toward the house. Marky jumped and twirled at my feet, thinking Hatsy had laid him a new ball. I let him sniff it and told him it was mine. His tail dropped. In the kitchen, Sarah and Danny gathered to view our miracle. They, too, beamed and cooed. It was amazing. It was simply an amazing egg.

I set Hatsy's awesome gift on a swatch of purple fabric on the kitchen table for all to view and appreciate. I took photos and e-mailed them to family and friends. The day after Hatsy laid that first lovely egg, she laid her second. The next day she laid the third, and she kept on going.

I had read that chickens take a day off now and then. Well, Hatsy didn't. There were even a couple of days when she laid two eggs. After the first 147 days, Hatsy did take a break, for about twenty-four hours. Then she went right back to laying an egg a day.

IN 4 MONTHS

this .... CREATED THiS.

It was kind of scary.

We hadn't been able to figure out what type of chicken Hatsy was, even when her adult plumage came in. But it was her awesome egg production that solved the mystery of her breed. Hers is known as a "production breed." Hatsy was a hybrid, designed for maximum efficiency. She was small and lean and was able to transform chicken feed into eggs at a dizzying pace. Hatsy's breed is the one preferred by factory farms. No wonder she had grown so quickly as a chick. Chicks that are quick to mature and early to lay can bring greater profit to farmers.

This also explained why she ended up so small and scruffy. In comparison to the other ladies, Hatsy looked downright scraggly. She was spending all her resources to make all those eggs, with nothing left over for elegant plumage. Danny and Sarah and I thought Hatsy was beautiful nonetheless, and we were proud of her fruitfulness.

We especially adored Hatsy's ragged bouquet of tail feathers.

A month later, Lucy and Lil'White began laying eggs too. Each girl's eggs were unique looking—something I never expected.

Hatsy's eggs were jumbo dark ovals. Lucy's were a lovely pointed pink. Lil'White's eggs were cute and perfect, just like she was, and she added white speckles to each one for decoration.

By the end of the summer, all three girls were laying full tilt. Our fridge brimmed with hard-boiled eggs, egg salad, and egg custard. We ate scrambled eggs for lunch, omelets for dinner, French toast for breakfast, and I baked cakes whether I wanted to or not.

Still, the eggs kept coming. I assembled six-packs as gifts for neighbors and friends, labeling each egg with the name of the hen who had laid it. I added a ribbon and a note to each carton. I wanted the recipients to appreciate these fresh home-grown eggs as the gifts that they were.

And they certainly did appreciate them.

And still the eggs kept coming. I bundled up a carton of eggs every time I left the house. I presented eggs to the librarian, the gas station guy, and complete strangers. I took pleasure in spreading the joy. I just hadn't expected this much joy from only three hens.

The girls and I settled into a nice daily rhythm. In the mornings they stayed in the coop and laid their eggs, and in the afternoons I put them into the tractor for a change of scenery. They got used to our routine and met me at the doorway to the coop, eager to be picked up and transported to greener pastures. I carried them one by one to the tractor.

One day I left the chicken tractor door ajar—just a crack, for just a minute—and Marky slipped inside.

7

Marky Joins the Flock

I dived into the tractor right behind Marky, groping and flailing in a blur of feathers and fur and squawks and fangs, but I couldn't catch him, so I shrieked. Deafening and furious.

And Marky stopped. Everything stopped. In the absolute silence that followed, a few downy feathers glided gently to the grass. I was crammed against the chicken wire inside the A-frame, hunched and shaking.

I glared at Marky.

His eyes glanced past me toward his escape route. In one fluid, desperate motion, he leapt over Hatsy, squeezed between my legs, and darted out the open doorway. He continued to run the length of our yard and stopped at the far corner. Glowing white in the deep shade of a tall pine, my little terrier guy turned to face me. He sat down.

I propped my shaking hands on my knees and looked around at the girls. I expected to see carnage, but there was no blood. Just some rumpled feathers and general chicken hysteria that settled pretty quickly.

I unsnagged my shirt from the chicken wire, backed out of the tractor, and latched the door behind me. The girls returned to their pecking and scratching business as if nothing had ever happened. What a resilient bunch.

I turned toward my dog.

Marky remained still. He was looking directly at me. And even from all the way across the yard, I could see that Marky understood. These chickens were mine, not his.

Marky was here long before the chickens. He'd been a member of our family for six years.

Danny and I had felt secure and happy with our decision to have a small family, but he and I had both grown up with siblings and dogs and chaos, and I thought it was important to throw a little wrench into the comfortably predictable lifestyle of our only child. When Sarah was about six years old, I felt that the time was right to add a family member in the form of a puppy. I wanted Sarah to experience the joy and companionship that I had enjoyed with my dogs, and I thought that there ought to be somebody around to occasionally gnaw the heads off her dolls.

For several weeks, I monitored local shelters on the Internet, waiting for our dream puppy to appear. When Marky's litter was dropped at a nearby shelter and I saw the photos online, I threw a laundry basket and a towel in the car and drove right over to take a look.

Oh, they were tiny. Eight weeks old. Part schnauzer, part Eskimo, they were a terrier mix. I asked the shelter lady if I could hold one. She opened the cage, picked up a male by his scruff, and handed him to me. And that was it. I had found Marky.

I didn't need the laundry basket and towel after all. He was so tiny and so quiet that I just zipped him up in my coat and there he stayed.

Back home with our new puppy, Sarah and Danny and I were ecstatic. Marky was adorable, and those first few days with him were heavenly sweet.

Then things started to heat up.

Marky turned out to be a take-charge kind of puppy. He stared me down. He bit every one of us, and not in a sweet puppy-chewing way but with the vicious intention of ripping our flesh.

We assumed this was just a phase. I barricaded him in the kitchen, as many people do with their puppies to keep them safe, although I was kind of doing it for our own safety. In his kitchen playland, Marky bypassed his squeaky toys in favor of the linoleum floor, prying it up with tiny fangs and peeling it off in sheets. Instead of gnawing his doggie treats, he carved the wooden legs of our kitchen chairs with the skill of a beaver.

Sarah took to crossing the kitchen by standing on one chair, then scooting another chair around and stepping onto it, then sliding the last chair around and so forth, eventually arriving at her destination without ever touching the floor, while Marky circled ominously below, snarling and spitting.

This was a little discouraging. I wanted my girl to adore her puppy, but Sarah's love for Marky was fading with every attack. My plan had gone wrong, and it was getting more horribly wrong by the day.

Concerned for our safety and for the safety of all who might come into contact with this tiny demon, I signed up for the first Puppy Preschool class I could find.

On our first night of class, people and pets mingled sweetly. I joined the circle and removed Marky's leash as instructed so that he could participate in joyful puppy play. Marky trotted into the fray, cute as a button. The first puppy he met was a poodle, and he greeted her with open jaws. She broke free and ran crying back to the lap of her human.

Marky went for the next puppy, and then the next. One by one, wounded, cowering puppies left the circle. I frantically pried Marky off his victims and shot apologies to their shocked owners.

Thank goodness our trainer was a take-charge type as well. She had been watching Marky, and she had a hunch. She picked him up and took a look at his teeth. Then she informed me that Marky was in fact not eight weeks old when he was dumped at the shelter; he was probably closer to five weeks. This is a significant issue because, she said, during weeks six, seven, and eight, the mother dog teaches her puppies how to socialize. It's during this period that, when a puppy nips mama's ear, she gives him a firm body slam. Dog mothers teach their pups clearly and quickly.

From that point forth, our trainer watched Marky with an eagle eye. When he launched himself at Daisy, an unwitting basset hound pup, our trainer hovered above, waiting for the precise moment when fangs met flesh. Then she snatched Marky's scruff, hoisted him above her head, and yelled at him.

For several days after that, Marky displayed perfect manners and total sweetness. My confidence in Marky grew as we attended class every week and practiced our homework. Marky and I became star students. As soon as Puppy Preschool ended, I signed up for Obedience School, and as soon as that ended, I signed up again. We worked on our commands every day, and after a good year or so of hard work, we had ourselves an awesome, trusting relationship.

Friends and family were amazed at Marky's transformation. Once unpredictable and downright scary, he had become mellow, well-rounded, and polite.

Marky and Sarah built themselves a nice relationship, too, complete with all the sweetness and the challenges I had hoped for.

Still shaking after our near disaster in the chicken tractor, I realized that this incident was my fault. The day he first met the chicks, I had decided that Marky was incapable of controlling his behavior. I had written him off completely as far as any chicken interactions were concerned. I didn't think he'd ever be able to change, so I didn't even try. Or maybe I didn't know how to start.

This day was the start. Marky was beside himself. Still under the tree, still watching me. I seized the opportunity.

We always begin with food. Treats and high voice and praise are the best tools. To help Marky understand his new role, I pulled a handful of chicken feed out of my pocket and offered him the first bite. Then I fed the ladies.

With this gesture, I assured Marky that he was the top dog, that he was the leader of my pack of chickens. He swallowed another handful of chicken feed for good measure, just to show the chickens that he pulled rank. Of course Marky's new leadership position meant nothing to the girls, but that didn't matter. Marky knew where he stood, and he was proud of his new job.

For a few weeks, Marky and I worked together on a short leash while the chickens free-ranged among us. We connected again, Marky and I, and our trust returned.

Marky took care to avert his eyes while he was among the girls, so as not to be incited by those eyeballs of theirs. A few years back, when we had our pet guinea pigs, Marky had displayed this same behavior as he attempted to maintain self-control.

Timmy | Cobbie

Timmy and Cobbie looked so tasty, and their eyes glistened like shiny black buttons, never blinking. To a dog, unblinking eye contact can be perceived as aggression. So Marky found that the best way to control his instincts in the presence of these creatures was to keep from ever looking at them. I was impressed with his self-control then, and now I could see that Marky was employing this same tactic with the chickens. What a good boy.

We began to work off leash among the ladies. The girls showed a bit of residual anxiety around Marky, but they eventually came to understand that he was probably not going to eat them. Still, they kept a safe distance and kept an eye on him at all times. They watched his stance.

If Marky was just milling about the yard, then the girls relaxed, too.

If he pricked his ears and trotted purposefully into the woods, Lucy and Hatsy expressed caution and stood tall to try to see what vicious predator had caught Marky's attention in the shadows. Lil'White, on the other hand, couldn't care less about danger. She was becoming quite the carefree chicken.

While Hatsy and Lucy did their free-ranging side by side, Lil'White wandered directly into the darkest depths of the forest, scratching and pecking on the fringe without a worry. Every now and then I rounded her up from the woods and returned her to the flock on the lawn, only to watch her saunter casually away again. Fortunately, Marky was vigilant. Foxes kept their distance, and Lil'White survived to see another devil-may-care day.

All the girls came to understand not only Marky's body language but also his vocalizations. To most of our ears, a dog bark is a dog bark. But the girls recognized the nuances of his voice. They paid little heed to Marky's "FedEx" and "mailman" bark, but they ran for the safety of the thicket at the sound of his "neighborhood dog coming for a visit" bark or "fox in the woods" bark.

Taking his new guard dog job very seriously, Marky set up a special morning routine. First, he strode out into the backyard, sniffed the breeze, and barked in the general direction of the woods,

just to let predators know that he was on duty. Then he patrolled the perimeter, peeing on choice trees and rocks at the edge of the yard.

When the girls free-ranged, Marky followed them from a comfortable distance while they meandered across the lawn. The girls occasionally dropped feathers on the ground, and Marky found them to be a delicious snack. A gift from the girls, just for Marky.

While the ladies and I were gardening one morning, the little white guy noticed that Lucy had a feather dangling from her wing, and he recognized it as a treat.

I diverted him with a command of "Leave it!" just before he plucked the feather, and just before Lucy would have plucked his eyeball in return. The girls were very tolerant of Marky, but they did have their limits.

Marky had his limits too. One day he was dozing among the lilies when a June bug landed on his tail. Marky didn't wake, but Hatsy sprang to action. She made a beeline for that shiny orange tidbit. A command of "Leave it!" wouldn't have stopped Hatsy, so I trampled the perennials to reach her just before she was to deliver the swift peck to Marky's tail that would have ensured her certain and untimely death.

While Lucy and Lil'White were cautious enough to stay away from Marky's things, Hatsy often felt the need to perform an up-close investigation.

And when I brought Marky his dinner bowl, Hatsy just assumed that the dog food was available on a first come, first served basis.

Marky disagreed. But his aggression never went beyond a growl, and that was sufficient to send Hatsy skittering away.

8

Chickenspeak

As I spent every free moment hanging out with the girls, I began to recognize their unique voices.

Hatsy was a sociable gal, and her voice was a classic clucking.

The way Lucy spoke reminded me of my grandma Alma. Grandma never had much to say, but she responded pleasantly to anything I would tell her. She'd offer a polite "Oooh?" or "Hmm" in order to keep the conversation going without adding much content.

Lucy even asked questions.

"Yes, Lucy," I replied.

"Brrrhrrr," Lucy said with a sigh.

Lil'White never spoke at all. No clucks, no chutters or exclamations.

But she sure could let out a good burp: a high-pitched ladylike belch that lifted her right off her pretty little feet.

I began to understand the meanings behind some of the ladies' vocalizations, especially the important sounds that warn of predators. When a hawk flew over, the first girl to spot it sounded the alarm with a high-pitched trill. Immediately, all heads rose and all motion stopped. They watched the sky and waited like little chicken statues until the hawk passed.

Hatsy tended to trill a lot of false alarms. But with experience, all three girls came to recognize hawks and to distinguish these very real dangers from sparrows, helicopters, and butterflies.

Marky had never really paid much attention to any of the

raptors or buzzards that circle overhead, and he didn't notice the girls' concern. To teach him that hawks are unwanted, I launched myself into an animated tizzy whenever I spotted one, pointing toward the sky and getting Marky all worked up. He quickly caught my drift, barked at the hawk, got a biscuit. Good boy.

Chickens have a different warning sound for earthbound predators. This sound is more like a growl than a trill, and is reserved for foxes, raccoons, and other fearsome creatures. My girls also have a specific "Look out, here comes Marky" expression that sounds like a subtle "oh-oh." Kind of like when kids spy the neighborhood troublemaker coming to join their game.

When a neighbor dog, Brody, wandered into our yard to play with Marky, I watched from the kitchen window to see what my guy would do. The chickens were safe in the coop, so I didn't have to worry about them. In the past, Marky would have invited Brody to play, and they'd have run laps across the lawn and ripped through the flower beds. This time, Marky barked and grumbled at Brody and physically blocked the entrance to our yard. Brody, disappointed and confused, turned and trotted home. I summoned Marky to the back door, and he came right over to receive his biscuit prize.

Of course, I later took Marky to Brody's house for a good playful romp on Brody's own turf.

A few days after the Brody event, I watched again from the window as Marky dashed barkingly toward the driveway. He had heard a loud rustling in the dried leaves in the forest and he voiced his most urgent warning, prepared to confront whatever monster might emerge. When an enormous tom turkey strutted out before him, Marky stopped and took a good look at it. They faced each other, and from the window I could see the cranks and pistons churning and chugging inside Marky's head. After a long pause he arrived at his conclusion.

The turkey was a chicken. Marky turned and padded casually back to his yard and the turkey continued on his way.

Marky was doing his job so well that I felt confident letting him watch over the girls all by himself as they free-ranged for short periods. One of those afternoons I let the girls out to roam the yard under Marky's watchful eye, then I went back inside to take the kettle off the stove. I was overcaffeinated and highly stressed that day because of an illustration deadline, and once I'd poured my umpteenth cup of coffee I forgot completely about the chickens. I rushed back up to the studio to finish my project. As I worked, I lost all track of time. Dinnertime came and went. The sun sank behind the trees. I never so much as looked up from the drawing board until I heard a quiet voice float through the open window:

"Brrrbit?" It was a chicken voice. Unmistakably Lucy's. "Ooooh. Rrrbrt," she moaned.

Aw. A sweet sound from the henhouse, I thought. Then it registered that the voice wasn't coming from the henhouse. It was right under my window. I put down my pens and went downstairs to investigate. When I flipped on the porch light and peered through the screen door, there stood Marky and Lucy. They were waiting patiently, side by side at the back door, for me to let them in. Far off in the darkness, barely illuminated by the porch light, I saw Lil' White and Hatsy huddled together at the closed coop door.

An hour earlier, when the sun had begun to set, all three hens had no doubt headed for the coop to get safely inside before nightfall. This is just what chickens do. Chickens tend to be very fearful of the dark, as they have almost no night vision. Unable to get into the locked coop, Lucy had left her panic-stricken friends and set off across the yard alone under the darkening sky to let me know it was bedtime.

That took thought. Maybe even a whole string of thoughts. And determination. And perhaps even reasoning. It was my first real glimpse into the tiny brain of a chicken, and I was profoundly impressed.

"Rrrrr bip," said Lucy.

How could I have left my flock out in the dark? A slipup like this could mean death to the girls. Raccoons, foxes, owls— so many predators come out at dusk. What a terrible chicken mom I was.

I scooped Lucy up in a football hold and carried her back to her flock while Marky trotted behind. I opened the coop, offered the girls my sincerest apologies, and planted them safely on their roost.

The next day I found an old roll of garden fencing and fashioned a run beside the coop. That way, the girls could be out, with access to their coop, and I could be forgetful without risking their lives.

I also used some of the fencing to make a couple of these handy chicken cages, to throw over the girls while they were out and about in case I needed to dash inside for a moment and then forgot the chickens for hours. Again.

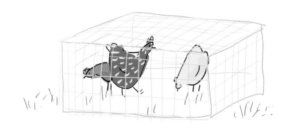

That summer I spent more time in the garden than I ever had. And when I started digging, the chickens came running. Grubs were their goal. Grubs drove them mad.

Whenever I set out to do a bit of gardening, I ended up grub hunting with the ladies all afternoon.

Patricia was right: these living lawn ornaments were fun and just about trouble-free. That summer with my girls brought more amusement and laughter to the backyard than I had ever hoped. The garden never looked better, Marky was thrilled that I was always outdoors, and Danny often joined me outside on weekend mornings for coffee with the girls. We had fresh eggs in the fridge and a healthy, happy flock out back.

Then, one morning in July, Lucy limped.

9

Lucy Limps

The problem seemed to be with her left foot. I guessed that she had just stepped on a splinter and it would work itself out on its own, but the next day she was still limping. Sarah asked me to call a vet, but I suggested we do our own examination. We picked Lucy up and gently flipped her over to see what we could see. But we couldn't find anything wrong with her toes or her legs.

The next day Lucy could barely walk, and she looked terrible. The crimson coloring in her face had turned gray, and for some reason her comb was curling over. While yesterday Sarah had been worried, today she was begging me to call a vet. But how could I? Lucy was a chicken. Marky's vet would laugh at me if I walked in carrying a chicken. You don't take a chicken to the vet the same way you don't take a goldfish to the vet. This was awful. I didn't know what to do. I told Sarah we'd keep an eye on Lucy and think about it.

By the afternoon, Lucy couldn't even sit up without balancing herself on her wings. Sarah was beside herself. I called Patricia to see if she knew what to do for Lucy. Patricia told me about Rosario, a farm animal vet who happened to live just up the road from us. I called Rosario right away, and despite it being dinnertime, she told me to bring Lucy on over. Sarah found a cardboard box, we tucked Lucy inside, and we hopped into the car.

Rosario met us out in her driveway. She lifted Lucy gently from the box and gave her a little exam on the front lawn. Lucy's wings hung low, her legs and feet were weak. Rosario explained that Lucy's problem was neurological, that it was possibly Marek's disease. There was nothing she could do for Lucy.

Marek's is a virus that affects young chickens and other barnyard fowl and can be fatal. It's an airborne virus. If one chicken comes down with the disease, then it's likely that the entire flock has been exposed to it. But since Hatsy and Lil'White were showing no signs of illness, they probably had a natural immunity. Rosario assured me that my family and Marky were in no danger—it's a bird disease and doesn't transfer to mammals.

To make us feel better, Rosario fed Lucy a syringe of garlic. She said it would help boost her immune system, and she handed me a few doses to take home. But she told us that all we could really do for Lucy was wait and watch, and that her odds were not good.

Back at home, since it was way past dinnertime, I automatically switched out of farmer mode and into mother mode. The box o' chicken was brought into the house and stuffed into the corner behind the kitchen table. Sarah started in on her homework while I set about cooking dinner.

Later in the evening as we were clearing away the dinner dishes, Sarah and I heard a soft voice.

We'd totally forgotten about Lucy. Sarah crawled into the corner and pulled out the box. We spread a red dish towel on the floor and arranged our beautiful sick chicken in the middle of it.

Lucy bupped again.

Sarah and I bupped too.

"Oooh," said Lucy.

Danny came home from work and sat down with us. We talked about Lucy's disease and prepared ourselves for the probability that she would not survive. We spent the rest of the evening sitting with her on the kitchen floor.

Lucy dozed as we chatted. Her face was so gray. She tipped to the left and used her wing to keep from toppling over. She quietly panted, with her beak open to cool her throat. Sarah fed her a couple of raisins and gave her a good-night pat, and we tucked our pretty hen into her cardboard box for the night.

The next morning, Sarah and I were up at the same time. I had hoped to check on Lucy first, alone, but Sarah was right there with me. On our way to the kitchen I reminded her that Lucy might not be with us anymore. I tried to slow Sarah down, but she scooted herself under the table and peeked cautiously over the edge of the box.

"Lucy?" Sarah whispered.

"Rrrrb," Lucy said. She wasn't quite dead yet.

We lifted her out of the box and placed her on a towel.
Sarah fed her, and then Lucy preened and slept.

Marky was very cordial with Lucy in his kitchen. He kept
his distance, only stretching his neck to give her a little sniff.
I don't know if he could sense that she was ill, or if he just
wasn't much interested in a chicken who didn't move.

It was a beautiful late-summer morning, and when I put
Hatsy and Lil'White into the chicken tractor for some grass
and sunshine, I placed Lucy in with them. She lay on the cool
lawn with bowls of food and water within reach, and seemed
comfortable with her friends.

I checked on her from time to time during the day, bring-
ing pieces of cold fruit and refilling her water bowl. That evening I put Hatsy and Lil'White back in the coop, but since Lucy
couldn't walk or sit or roost, I had no choice but to bring her
back into the house. I sat down on the kitchen floor with Lucy
on my lap, pried open her beak, and squirted one of Rosario's
syringes into her mouth. She shook her head and spattered us
both with essence of garlic.

Then I helped her get comfortable on the red dish towel and she watched me prepare dinner.

"Herrrbert," said Lucy.

I told her what I was cooking.

"oooOOOooh," she said.

"Oh?" I asked.

"Bup," said Lucy.

The next day it was raining, so Lucy stayed indoors with me. I set her up with food and water, and I dug up one of the tiny training roosts I had built for the girls when they were chicks.

Lucy seemed more comfortable when she had something to wrap her toes around. She couldn't stand up, but she could balance on the roost, and since it was only a few inches off the ground, she wouldn't hurt herself if she tipped over. This kept her from sitting in her poo, too. With a newspaper on the floor under the perch, we had a pretty tidy setup.

Days passed. Lucy lived.

She spent a lot of time sleeping.

When she was awake, she kept me company in the kitchen. I moved my art supplies down from the studio to the kitchen table so we could hang out together while I worked.

Lucy did make lovely company.

She also looked quite stunning on the white futon.

In the afternoons, she and Sarah watched television together. We were kind of having fun with our house chicken. She was friendly, she was polite, and she never moved from wherever we placed her.

But as much as we enjoyed her company, I could see that Lucy missed her flock. I saw her perk up when their voices floated in through the open window.

She could see them from the futon, and she watched them.

So on nice days, I carried Lucy out to spend time with Hatsy and Lil' White.

Lucy's preferred mode of transportation: Sarah's Easter basket. I kept the three ladies together in the tractor instead of allowing them to free-range around the yard, so Lucy wouldn't be left all alone if the other two wandered.

When she was out there with her friends, Lucy sat a little taller and tried a little harder to look like a healthy bird.

Sometimes she managed to stand for a moment or two.

When I brought Lucy out, Hatsy was always waiting to

greet her friend. Then the two of them would get to work picking clover together.

Lil'White would recede to the far end of the tractor, and Lucy would fit herself back into second place in the pecking order for the day.

While I worked on my illustrations, I kept a close eye on things through the kitchen window. Lucy tended to scoot backward away from the food and water, so from time to time I went out to check on her and rearrange things so she could reach what she needed.

On my way back to the kitchen, Marky greeted me and invited me to play, so I did. Then the garden beckoned me to weed it, and in no time at all the day was over and I had gotten a nice tan.

When deadlines loomed, however, I stayed on task. One such day, after hours spent hunched over my laptop, I lifted my head and stretched. Out in the tractor I saw Lil'White pecking around, but Hatsy was sitting down beside Lucy. She wasn't digging or grub hunting. She wasn't moving. This was very odd. Hatsy never, ever stopped moving. She was in perpetual motion from dawn to dusk. My immediate thought: Hatsy had come down with Marek's.

I went to take a closer look. Hatsy was still down as I approached the tractor. But when I got closer, my little orange dynamo popped to her feet and dashed over to say "hello!" She was fine. She had been sitting on the grass only to keep her friend Lucy company.

I made up my mind at that moment that Lucy should no longer be treated as a house chicken. She was important to her flock, and she should be with them as often as possible.

I removed Lucy's box from the kitchen and set up a new home for her in Marky's old dog crate on the screened porch. Each day she would spend as much time as possible with the girls, rain or shine, and she would spend nights safely outside. She took to her new home just fine. On the porch, Lucy enjoyed cool breezes and the sounds of her friends from across the yard. After a couple of weeks, she started laying eggs again. That was certainly a good sign, but it was about the only one. Her legs were still very weak, and her toes were stiff and bent. A bit of crimson coloring had returned to her face and comb, but she was still sleeping more than she was awake.

Out on the porch on warm summer evenings, Lucy perched on my knee and joined Danny and me in some pleasant conversations. But with autumn around the corner, I began to wonder what I was going to do with her. Would she ever be able to move back in with the flock? Was I destined to care for a paralyzed chicken indefinitely?

I found myself sketching tiny wheelchairs.

10

Change in Pecking Order

Now I could see why three or more chickens constitute a flock. When the three girls were together in the tractor, there were lively interactions and discussions, plenty of activity, and a hint of conflict to make it all interesting. When one gal discovered a grub, someone else surely came running. A mosquito flitted past, and again a flurry of activity followed.

Lucy's absence from the coop left a big hole. When Hatsy and Lil'White were together, they seemed to be just hanging out. The two carried on fine in the coop, but it was as if they were living in slow motion.

I continued to take Lucy out to be with the girls every day so that they could be a flock for at least a couple of hours. Once she was placed in the tractor, things bustled again. Lil'White and Hatsy busied themselves. And even though Lucy didn't move far from where I put her, it seemed that each day she sat a little taller in the grass.

In August, the opportunity arose for our family to go to southern Spain.

Danny and Sarah and I had never taken a vacation abroad together, and we were very excited. I arranged for Marky to stay with friends. Sarah's Betta fish, Betsey, moved in with the neighbors. My chickens—well, that would be tough. How could I ask anyone to care for a paralyzed chicken? Hatsy and Lil'White could pretty much manage for themselves in the coop, but Lucy needed quite a bit of attention. Her food and water had to be kept fresh and within her reach. Poops had to be scooped from her crate, and her eggs had to be collected. I made a couple of phone calls and was surprised and flattered when two wonderful friends stepped right up to help.

My friend Beth loved my girls. She had come over to our house to play with them when they were wee fluffy nubbins, and she happened to have a flock of her own, of sorts.

Beth had four ducklings swimming around in a wading pool in her backyard. She loved animals of all kinds and was very happy to help out with Lucy's care.

Our neighbor Trish, who is a nurse, a massage therapist, and a whole list of other things, also agreed to help out. I showed Beth and Trish the ropes and organized a schedule for them, and then I took off for Spain with my family.

I'd like to say that I didn't worry about my Lucy during that vacation, but at best I tried not to obsess about her. If she didn't survive my absence, well . . . I surely had done the best I could for her.

Ten days later, back in the States, back in Upton, back on the back porch, I greeted Lucy with a big warm smile. And Lucy stood up to greet me.

She stood right up!

I opened the cage and she lunged awkwardly toward my arms. I lifted her and carried her to the lawn. As soon as her feet touched the earth, Lucy stood and took a few pained steps. Then she looked up and so did I, to see Trish striding up the path from our driveway. Trish planted herself on the grass beside us and told me how much she'd enjoyed spending time with Lucy. She explained that twice a day she had taken Lucy out of her cage for chicken therapy, holding her up, helping to stretch her legs.

This was something I had never thought to do and something wonderfully above and beyond. She told me that while

I was away, she and Beth had met each other in the backyard during one of these therapy sessions, and Trish taught Beth the chicken therapy moves. So Lucy had been through a rigorous schedule of exercise and affection, two and three times a day, for ten days. And it showed. I watched Trish's eyes as she exclaimed how dear a creature Lucy was.

While the therapy and Lucy's improvement were thrilling to me, I was even more stunned to hear from Trish how she and Lucy had bonded. This was a solid reassurance that I was in fact not totally insane.

I knew that I had fallen in love with a chicken. Now I knew I wasn't the only one who had.

Out in the coop, Hatsy and Lil'White seemed happy and well cared for, too. Later that day we retrieved Marky from our friends' house and Sarah's fish from our neighbors across the street, and we all settled comfortably back together.

The next morning when I opened up the coop, Hatsy and Lil'White tore out the door like a pair of escaping convicts. All memories of their ten tedious days in lockdown evaporated, and they set to work ripping up the lawn. I rounded them up and placed them in the tractor on a cool patch of overgrown grass.

Then I brought Lucy out to join them. She staggered around a bit and sat down to graze. Soon Hatsy was at work on a new tunneling project in the corner of the tractor. Lil'White sauntered about and paused, standing over Lucy.

And then Lil'White snapped.

She struck Lucy with her beak. *Whack,* right on the comb. I winced. Lucy ducked. *Whack.* Fierce and precise. *Whack.* Lucy struggled to stand, but Lil'White was in the way. *Whack.* Lucy sank down onto her haunches and lowered her head. The attack continued, rhythmically, methodically. I reached in and brushed the vicious Orpington out of the way. Lucy lurched toward my hands and I helped her out of the tractor.

I put Lucy down and kneeled on the lawn beside her. A bubble of dark blood oozed from her ripped comb. I looked back at Lil'White. She pecked daintily at a piece of clover as if nothing had happened. Hatsy was still digging. Her pit was now formidable, and she retained perfect focus. I stayed with Lucy for a while and then returned her to the crate on the porch.

I thought that Lil'White might just need some time to adjust after those ten days without Lucy, so the next day I tried to integrate them again. Immediately Lil'White attacked. I removed Lucy again and placed her outside where Lil'White couldn't reach her. But Lucy's head remained low.

Lil'White had made herself clear.

I was shocked and horrified by Lil'White's behavior, but I wasn't angry at her. Chickens will peck at an injured or sick flock member, sometimes to the death. They have wild roots, these girls, traced back to the jungle fowl of Indonesia. So while my girls are domesticated animals, life in a coop is not a natural existence for them. Domestic chickens get along well only if they have plenty of space to get away from each other. Chicken keepers know that vigilance is key. If a bird is injured, the wounds must be addressed promptly to reduce risk of further injury from the flock.

I did my best to understand Lil'White's behavior. I supposed she had been acting on her instincts. In the wild, a sick bird like Lucy could attract predators and put the whole flock at risk, so it might be best if Lucy were not around. Lil'White was still as kind as ever to Hatsy.

Lil'White's charming behavior didn't change toward Danny or Sarah, either. Lil'White and Danny still shared a special rapport. And Sarah refused to see any quality but sweetness in her favorite chicken.

But I have my reservations about Lil'White's supposed charm. Because I am a victim of Lil'White, too. Even now, whenever I enter the coop, Lil'White brutally attacks *me* every chance she gets, appearing innocent and curious all the while.

My legs receive most of her abuse, but whenever I squat down to chat with the ladies, Lil'White quietly slips around back and pecks mercilessly, maniacally, at my rear end.

I believe that if she had the appropriate weapon, she'd kill me, drag me into the woods, and bury me in a shallow grave behind the compost pile. But all she has is a little beak and a twisted mind. She doesn't scare me one bit. She's a wonderful beautiful mystery, Lil'White. But she's also completely off in the head.

Inside the chicken tractor, Hatsy had witnessed that second attack on Lucy, but she hadn't joined in. She was just as kind toward Lucy as ever, and her pleasant relationship with Lil'White remained intact as well.

I made a slight change to our daily routine by setting Lucy out in a cage beside the tractor. Sometimes Hatsy stayed with Lil'White, and sometimes she grazed in the cage with Lucy.

I was still devastated by Lil'White's brutal gesture, but Lucy took it in stride and adjusted to the new setup just fine. She was getting stronger, and I stopped worrying whether or not she would survive. But she still hobbled and limped on those bent toes, and walking and climbing the ramp were a challenge for her.

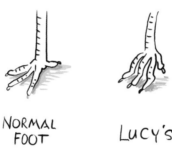

NORMAL FOOT   LUCY'S

It was apparent that her feet had been damaged by the Marek's disease, and they weren't healing.

I wondered again about Lucy, about autumn and beyond. She was going to need a real home. She was going to need a special-needs coop.

11

Wintertime

I designed a compact coop for Lucy.

The footprint was about four feet square. Its construction was pretty simple, though it was just as painful to put together as the big coop thanks to that pesky hardware cloth. In the end a nice coat of paint hid my bloodstains on the wood, and from a distance the coop looked just darling.

I wanted Lucy to live as near to her flock mates as possible, but there is a really nice boulder at the back edge of the yard that has always been a scenic focal point in winter, so I placed her coop to the right of that rock. Carefully placed pots of geraniums completed the bucolic look.

In her new home, Lucy had every comfort.

She had a nice log to perch on, and I hung food and water in recycled yogurt cups all around the edges, upstairs and down. There was even an open-air balcony for warm nights. No matter where she was, Lucy would always have a clear view of her flock in the big coop. I gave her a window that faced our own house too, since I knew that all the girls enjoyed watching us as much as we enjoyed watching them. In the back of the coop, a multitude of doors ensured that I could reach Lucy easily, just in case she needed me.

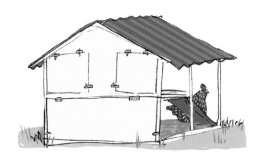

On these crisp autumn days, the three girls enjoyed spending time together and digging in the leaves at the edge of the yard. Lucy enjoyed being one of the flock at these times. She was safe from Lil'White's aggression as long as Lil'White was busy bug hunting.

The cooler weather seemed to give Lucy a boost of good health. The summer's heat had been a challenge for her, especially since she slept on warm pine shavings instead of on a cool roost. On the hottest days, Lucy panted like a dog and spread her vast wings to cool her body. She was by far the largest of the three hens, and while she wasn't as fluffy as Lil'White, she did have a voluminous coat of downy feathers. Now during the cool autumn days, Lucy was more animated and even mustered the energy to help Hatsy with her digging.

When Lucy's energy ran out, she sat down and became the watch chicken, scanning the skies for danger while Hatsy strip-mined the forest floor.

Eventually Hatsy and Lil'White moseyed to the next leaf pile, leaving Lucy behind. Too exhausted to stand, Lucy could only sit there and watch them go.

It was an exercise in patience.

Lucy waited as long as it took for me to happen by and transport her to the next locale.

Sometimes I put Lucy and Hatsy together in a cage on the lawn so that Lucy could feel safe and included.

Hatsy was perfectly contented hanging out with Lucy. Anywhere she could dig a pit, Hatsy was a happy gal.

Nearby, wicked Lil'White enjoyed a cage of her very own.

While the girls genuinely delighted in the cool days and nights of autumn, I wasn't too sure about their ability to keep warm in the frigid winter. In the big coop, I always left the door to their sleeping quarters open because the food and water were down below, and the ladies enjoyed coming and going as they pleased within their safe confines. The entrance to Lucy's nest box had no door at all, for the same reason.

But on the most blustery winter nights, as I huddled in bed under my own warm comforter, I imagined the girls in their coops, shivering and miserable. When the weather report predicted a cold snap with nightly temperatures dipping into the teens, I rigged up cozy heat lamps in the sleeping quarters of both coops.

That night, all three of my ladies chose to sleep out in the cold rather than in the warmth of those glowing lamps, and I felt like a fool. I unrigged the heat lamps the very next day.

When the first snow fell, I shoveled a path to the coops and opened the doors to the girls' new experience.

Lil'White took a flying leap

and landed, stranded. She didn't move. Didn't look around, didn't struggle. She waited. For the thaw, for the rescue, for whatever might come.

Of course I rescued my golden girl from the snowdrift and placed her on an old pallet from which she could observe her beautiful white world.

On really cold days, Lil'White looks more like a basketball than a chicken.
Her secret:

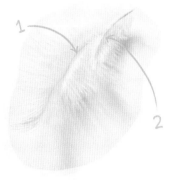

Double-shafted feathers. Double the insulation.
Makes me feel all cozy just thinking about it.

Her Orpington pantaloons are extra fluffy in winter to protect those dainty ankles.

She even puffs her head feathers to protect that petite comb.

Every freezing morning, I carried a teakettle of hot water to the coop to thaw the ice in the water bowls. When I had the time, I let the girls out to explore their white wonderland.

Lucy was looking prettier and healthier every day.

But the cold and the snow didn't appeal to Hatsy like they did to Lucy.

Hatsy preferred to perch high above it. Here on her perch she hunted for shiny things to peck at: buttons, zippers, eyeballs . . .

Hatsy was truly impressed with Lucy's private coop.

Whenever the opportunity presented itself, she rushed in and took a thorough tour. She sipped Lucy's water, she sampled the multiple cups of feed, she tried out the nest box. She inspected every corner and crevice as if she were hunting for Lucy's secret stash. When her investigation was complete, she sat down on the log beside Lucy and the two friends preened and conversed.

Sometimes when I closed up the coops, Hatsy refused to leave Lucy's, so I allowed her to stay and visit for the day. Other times she stayed for a sleepover.

As dusk fell, Hatsy squeezed in beside her oversized friend on the open balcony, and Lucy's voluptuous bulk kept them both warm through the frigid night.

Hatsy's visits helped break the monotony of winter.

But for the most part it was a lonely time.

Winter is just not the easiest season, for any of us.

I couldn't let the girls out unsupervised, even with the fencing I had put up, because Lil'White tended to corner Lucy and peck the living daylights out of her.

I left Lil'White locked up sometimes, so Lucy could relax and enjoy her free time with Hatsy.

When I had the time, I brought my sketchbook and pulled up a stool and hung out with the flock.

Lucy bumped deliberately into my leg to let me know I should pick her up.

When Lucy was sitting on my knee a safe distance above Lil'White, she loosened up and launched into animated conversation. She and I had some nice discussions while I sketched. I wish I knew what we were talking about.

Lucy's plumage was breathtaking against the white snow. Bright red coloring had returned to her face and wattles, and her eyes sparkled. I could tell that she felt healthier, and I was so pleased. I only wished Lil'White would change her mind and allow Lucy back into the flock.

Lucy, however, didn't appear to harbor such wishes. Or any wishes. I wished that I could "live in the moment" as masterfully as Lucy did.

Marky is pretty good at living in the moment, too, as long as the moment doesn't include thunder and lightning.

He also loves the cold weather as much as Lucy does.

I gave my staunch outdoorsman this fancy doghouse, but he never entered it willingly.

He prefers to face the elements head on.

Marky's a snow dog, and that's that.

So the doghouse went to the chickens.

Marky didn't miss it at all, and the ladies appreciated their new diversion.

One afternoon I found Hatsy tucked into the corner of the doghouse. I thought she might be trying to lay an egg in there, but then I watched her eyes begin to close, and her head drooped forward. This was surprising; I had never seen her so subdued.

When Lucy hobbled in and stood beside her friend, I thought that something must be wrong. Lil'White poked her head in too, took a quick look and moved on. But Lucy stayed with Hatsy. When Hatsy's head tipped so far forward that her beak rested in the bedding in front of her, I knew for sure that Hatsy was ill. I thought about pulling her out for a closer look, but she seemed to be in good company beside her big speckled friend, so I decided not to interfere. I left her in Lucy's care and came back to check on them from time to time. It took a couple of hours for Hatsy to snap out of her trance, but when she did, she and Lucy returned to their chicken pursuits like nothing had ever happened. I was puzzled but relieved.

The next morning Hatsy greeted me with all her usual vigor and spunk. When I checked the nest box for the morning's gifts, I discovered a fresh Lucy egg, a fresh Lil'White egg, and a third, frightfully enormous, pterodactyl egg.

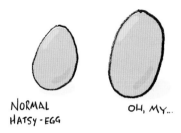

NORMAL
HATSY-EGG

OH, MY...

That one had to be Hatsy's. I guessed that its horrific size must have been the cause of her discomfort, and I sincerely hoped she wouldn't be popping out any more of those monstrosities. When I brought the eggs to the kitchen, I took a last good look at the dino-egg and chucked it into the trash. That egg represented Hatsy's pain, and I just didn't want to know what was inside.

After that horrific egg appeared, Hatsy continued on her sprightly way, and weeks went by without incident. The flock and I moved happily through the season as if it had never happened.

Layers of fresh snow blan-
keted the yard in clean, cold
loveliness and the girls' world
shrank to just their two coops
and a small path between
them that I kept shoveled and
trampled. There were stretches
of bad weather when I just
couldn't let the girls outside
at all. To break the monotony,
I hung piñatas of apples and
broccoli in the coops.

Other times I hid sunflower seeds and peanuts, which
kicked off an animated treasure hunt. And on mornings when
I cooked oatmeal for my family, I made extra for the girls.

Occasionally Lucy came into our house for a visit.

She wandered comfortably from room to room, checking the
place out and offering the occasional comment.

I had read about "chicken diapers" for house chickens and decided to try one out on Lucy. I found a pattern online and stitched her a pretty blue one.

Once I had wrestled her into it, the diaper was nearly invisible beneath her feathers.

But she carried on as if I had given her a horrible wedgie, so I took it off.

We kept our house visits pretty brief, so poop wasn't an issue anyway.

I loved our one-on-one time. I loved that Lucy seemed to enjoy it too. She's the one who initiated our social engagements. Sometimes when I opened the door to her coop, she just looked at me and bupped, unwilling to move or even to chat much. On those days I gave her a raisin, shut the door, and that was that. But when Lucy wanted to get out, she made it known. She tottered as quickly as she could through the open doorway, greeting me with a moan, "Errrrrrb," and bumping her chest into my shin. I bent down and offered her my arm and she stepped onto it, and then we were on our way. Lucy and Lauren.

Chicken and her vehicle.

Sometimes I fancy myself a falconer—my regal Barred Rock clenching the gauntlet with mighty talons of death . . .

Lucy enjoyed being my muse. She served obligingly as an artist's model and proved capable of holding a nice long pose while I sketched.

We launched other creative exploits together, and as long as they didn't involve chicken diapers, she was usually happy to work with me. One day I set up a photo shoot for the fun of it.

She humored me for a while,

then she walked off the set.

Spring arrived none too soon.

The snow melted, the warm sun shone, and the ground thawed.

Hatsy's rototilling drive kicked in,

and our yard once again became a field of small perilous pits.

Lil'White sauntered around the garden beds in search of one particular delicacy:

plastic plant tags, the skinny labels that poke out of a flat of seedlings from the nursery. Whenever she found one of these tags, she picked it up and whacked it on the ground, attempting to break it into bite-sized pieces. I followed her around and snatched them out of her beak before she could swallow them and die. Each time I grabbed a tag, she moved on to discover another. I didn't recall leaving these tags all over the garden, but Lil'White sure dredged up a passel of them.

While Lil'White and Hatsy frolicked joyfully in the bright spring sunshine, Lucy wanted none of it. I had a heck of a time prying her out of the shadows of her coop. She just did not want to leave the nest box. I grasped her firmly and lifted her out and shut the door so she couldn't get back in.

Out on the lush lawn, Lucy shook herself indignantly and puffed out like a speckled balloon. She paced back and forth beside her closed door, head down, muttering to herself. She was in quite a state.

Since Lucy wasn't about to change her tune, I gave up and allowed her back inside. She hobbled straight up the ramp and plunked herself back down on the nest. After a couple days of this same drill, Lucy's muttering morphed into a chant: "Budup . . . budup . . . budup . . ." This was a word I had never heard before, from Lucy or any of the girls. And once this chanting began, it didn't stop. "Budup" was Lucy's mantra, and it continued all day and all night.

I wasn't worried about her. I knew Lucy hadn't gone insane. She had gone broody.

12

Budup

Lucy wanted to be a mother.

It was a hormonal thing. Broodiness happens. I was excited to be witnessing something I had read so much about. But I was disappointed that Lucy was expressing such strong commitment to a desire that could not be fulfilled.

We didn't have a rooster. And since we had no rooster, Lucy's eggs were not fertile. And Lucy had stopped laying eggs, anyway, because that's what happens when a hen goes broody. I had absconded with every one of her eggs, day after day, so there weren't any left for her to sit on, fertile or not. This small detail, however, didn't make any difference to Lucy. All she wanted to do was brood, eggs or no eggs. She would have sat on golf balls if I had given her some, but imaginary eggs were fine too.

Budup.

I could relate. Years ago I had also desperately wanted to start a family. I brooded. I nested.

Then our girl Sarah came along, and Danny and I were both overcome with parental joy. We still are. Well, sometimes.

Some breeds of chickens are more likely to go broody than others. Throughout history, quite possibly as many as ten thousand years of poultry-keeping history, humans have carefully selected for and against certain characteristics in their chickens. The tendency to go broody is often considered undesirable, especially for farmers who count on their hens' high egg yields. These farmers appreciate a chicken who steps into the nest box, drops her egg, and walks away without looking back. This makes the eggs easy to collect. It also ensures that the hen will keep on producing.

A hen who is ready to set, or brood, stops producing once she has accumulated a number of eggs—a clutch. The broody hen then stays on the nest for approximately twenty-one days, the amount of time it takes for her chicks to develop and hatch. During this brooding period, she leaves the nest only two or three times a day—to eat, drink, and poop.

Farmers have devised all sorts of strategies to break the spell of a broody hen and get her back into production. Sometimes just locking the hen out of the coop for a day or two is all it takes. Some people swear by the wet-hen method: dunk that broody in a bucket of cold water. For tougher cases of broodiness, some farmers place the hen in a wire-bottomed cage, along with food and water. The cage is suspended on blocks above the floor for three or four days. Air circulating under the hen cools off her belly and encourages her system to resume laying. Another tried-and-true technique: frozen peas.

At least with the frozen-pea method, the whole flock can celebrate in the end with a feast of all those thawed peas.

None of these broody-breaking strategies appealed to me. I certainly didn't want to do anything to make Lucy uncomfortable—she was experiencing enough discomfort already with those bent toes and weak legs. And broodiness isn't necessarily an awful thing. She wanted to be a mother. Who was I to stop her?

I considered my choices, I thought about Lucy's challenges, and I made my decision. We would embark on a new adventure, Lucy and I.

I discussed my thoughts with the indoor contingent. Sarah was intrigued with the prospect of adding some new chicks to the flock. Danny was delighted with my promise that they would not be raised in our living room.

I picked up the phone and once again dialed my partner in chicken exploits. Patricia had a rooster in her flock, so the eggs produced by her hens were likely to be fertile. Her rooster was a fancy little guy. Colorful, with white earlobes. Patricia had adopted him and didn't know anything about his background or what breed he was. This proud little rooster didn't crow often, but when he did, he puffed up his chest and mustered all his vocal strength to belt out a sweet miniature yodel. He was a friendly fella, too, which certainly is not something you can say about all roosters.

And he was a favorite of Christopher, Patricia's son.

At Patricia's invitation, I drove on over the next morning to collect some fresh fertile eggs from her gals. It was a cold day outside but in the henhouse it was warm and cozy. In a bank of nest boxes on the far wall sat two lovely puffed-out hens who paid us absolutely no mind. Patricia reached under a full-figured Rhode Island Red and pulled out two eggs. If the hen noticed, she didn't show it.

I wrapped the two warm eggs in a couple of tissues and slipped them into my coat pocket. Patricia suggested that I take a few more eggs but I didn't want to overwhelm Lucy or myself. She assured me that I could always come back for more if these two didn't hatch. I thanked her and hurried home.

The eggs were still plenty warm when I presented them to Lucy. With a calm matter-of-factness, she lifted herself slightly and then tenderly guided the eggs with her beak to just the right spot underneath her. Then she sat down.

That was it. No "Thank you," no nothing.

That was okay with me. I latched the door and left her alone to do her thing.

The next day I let Lucy stay on her eggs rather than boot her out onto the lawn with the girls. Hatsy and Lil'White sensed that something was up, and they curiously milled about her coop.

Since Lucy was going to be eating only one or two small meals a day while she brooded, I prepared for her an extra healthy protein-rich concoction. The primary ingredient: a scrambled egg. This did feel a bit odd to me—feeding a chicken egg to a chicken. But it is known among chicken folk that eggs are one of the healthiest foods to give a chicken. After all, that one yolk is packed with all the nutrition it takes to grow an entire chick from scratch. I added a sprinkle of oatmeal and some fruit and greens to the mix and brought it out to Lucy's coop.

I opened the door and Lucy turned to look at me. She was very solidly planted on that nest, and I was not looking forward to wrestling her out of there again. So I decided to offer her the food and water right where she sat.

Lucy ignored it. She didn't want breakfast in bed. I placed the bowls of food and water on the ground, rolled up my sleeves, and geared up for a struggle as I reached in to collect her. This time, however, she gave me no resistance. I think she was probably pretty hungry. And maybe after my gift of fertile eggs, she perceived me to be an assistant rather than a tormentor. Out in the rain-soaked yard, I placed Lucy in front of her bowl.

She ate ravenously and then took a long drink of water. She gave her plumage a good shake and looked around, then wandered off to graze for a while. I watched her go and wondered if she would be keeping track of the time. After all, it was a chilly, wet day. How long could a couple of eggs wait, and how cold could they get? Lucy kept on walking.

All along, Hatsy had been watching. I didn't allow her to share Lucy's special meal, so she hovered around the little coop until Lucy finished eating and walked away.

This was the moment Hatsy had been waiting for.

As soon as Lucy disappeared around the corner, Hatsy made a beeline for the open door. I was surprised by her speed and stealth and almost reached out to shoo her away, but then I stopped myself. I wanted to see what she would do.

She hurried up the ramp and peeked into the nest box.

She approached with delicate footsteps and uttered a few sweet sounds.

Using her beak, she scooted the eggs around a bit. Then she sat down.

Hatsy tried to cover the eggs with her bony little breast, but no matter how she rearranged them, one egg kept popping out. She spoke to the eggs. The words she used were pleasant and hushed.

Hatsy was able to enjoy only a few blissful minutes on the nest before Lucy's behemoth mass came tottering back up the ramp and filled the doorway. Lucy looked down at Hatsy. And Hatsy promptly surrendered the treasures and launched herself toward me through the open door. I ducked away from her beating wings, and she landed on the grass beside me. She turned to watch her friend.

Lucy hobbled on over to the nest. She angled herself into position and then sank gently onto the eggs, resuming that rhythmic chant, "Budup. Budup."

I closed Lucy's door.

Hatsy walked away.

13

Team Broody

Nineteen days to go. Nineteen days to watch and study and obsess and plan and hope and worry.

When I wasn't outside snooping around Lucy's coop, I was inside snooping around the Internet—reading up on any and all chick-hatching topics. I was surprised at how much there was to learn. Sarah and Danny enjoyed hearing me expound about egg development and such, but rather than join me in my studies, they were satisfied to receive brief updates on Hatsy and Lucy's activities and to hear the occasional fun fact about broody hens.

One night I cornered Sarah and read to her about candling. She enthusiastically agreed to join me in this experiment. We read that after the seventh day of incubation you're supposed to be able to see the beginnings of a chick inside the shell by shining a bright light behind it. Of course we couldn't wait seven days. So on the fifth night, Sarah and I went out and collected our precious eggs from under Lucy and brought them inside. We sat on the basement steps in total darkness fumbling with flashlights and eggs.

When we held the first egg up to the flashlight, it glowed with a uniform golden hue. It was pretty, but there was nothing interesting going on in there. In the other egg, however, we very certainly saw a web of veins.

This egg was fertile! Sarah and I discussed what should be done with the one that was infertile. I thought we should dispose of it, but Sarah felt Lucy might get upset if it mysteriously disappeared. I agreed that there was no harm in letting her continue to sit on both eggs. With a pen I drew an X on the infertile one so we could tell them apart, and then we returned both eggs to Lucy's coop and tucked them under her wonderful fluff. "Budup. Budup. Budup." What a nice sound out there in the dark, warm night.

Sixteen days to go.

Lucy continued to sit faithfully on the nest, but all that sitting took a toll on her legs. They got weaker and weaker, and eventually she couldn't even lift herself to stand. But I was as committed to our broody adventure as Lucy was, and I was there for her when she needed me. I kept a regular schedule that she could depend on: two or three times a day I lifted her off the nest for a leg stretch and a nice meal.

I remembered how my friends Trish and Beth had helped Lucy to stand, and I tried to do the same. It took awhile, but as I held her, hovering just above the ground, she was able to stretch her legs and then put weight on them.

Throughout our broody adventure, I remained wary that one day Lucy might turn on me. I had read so many accounts of the maniacal demeanor of broody hens. Broodies can be downright nasty, hissing and pecking at anyone who invades their space. But Lucy never did go nutty on me. She was polite and trusting and patient, always. She needed me. We were a team.

Auntie Hatsy was proving to be a valuable team member, too. Each time I took Lucy off the nest, Hatsy shot right in and plunked herself gently on top of the eggs . . . more or less. Lucy seemed perfectly confident with Hatsy's babysitting skills.

I was pleased with the girls' tag-team setup because I could tell that Lucy really appreciated her free time on the lawn.

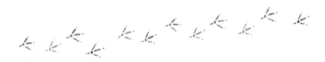

There was really only one unpleasant by-product of our broody adventure. But it was a doozy.

THE BROODY POOP

Yes, I had read accounts of this broody poop. But nothing I read could have prepared me for the real thing. It emerged right after Lucy finished her meal, and it was positively the most vile excrement under the sun: bigger than a golf ball, and it absolutely reeked. Regular chicken poop doesn't smell like roses either, but this broody poop was a collection of half a day's worth of waste, festering inside the bird until she stepped off the nest.

I kept a long-handled shovel nearby with which to scoop the fetid turd and hustle it to the compost pile, where I buried it deep.

Inside the nest box, Lucy kept everything remarkably pristine and tidy. She didn't poop in there, and she used her clean beak, rather than her toes, to turn and arrange the eggs just so.

When Lucy was away, her friend Hatsy did some rearranging of her own, but that didn't seem to bother the lady of the house. Upon Lucy's return, Hatsy always gave up the eggs without dispute. Lucy stepped in and moved everything back into its proper place and resumed where she had left off.

Sometimes Hatsy had a hard time leaving.

After a while, Hatsy's love for Lucy's eggs began to verge on obsession,

and Lucy seemed a bit annoyed by her persistent little friend.

Meanwhile, Lucy graciously tolerated my own annoyingly frequent attention. I spent more and more time with Lucy and the flock. And in the evenings I spent more and more time on the Internet, learning whatever I could about everything chicken.

One night while cruising the chicken-lovin' websites, I read that an infertile egg left on the nest could actually explode.

I grabbed a flashlight and scurried right out to Lucy's coop and removed the festering time bomb.

She never missed it.

I read a gripping account of egg production within the hen.

Eggs are formed one after another, conveyor-belt style. It takes about twenty-five to twenty-eight hours to create one complete egg. At the end of the process, the shell is applied, and then a special water-soluble antibacterial coating called *bloom* is the finishing touch. This coating protects the porous shell and its contents. It's the reason that freshly laid eggs, laid in a clean nest box, need not be washed. It's also the reason that fresh unwashed eggs can sit in a bowl on the kitchen table, unrefrigerated, for several days without spoiling. Eggs sold in supermarkets are generally produced on factory farms. The bloom is washed off before they are shipped, so factory eggs do need to be refrigerated from the start.

The egg's shell is composed mostly of calcium, with a slight bit of protein to bond it all together. A laying hen needs to consume a good quantity of calcium in order to create each of these shells, day in and day out. That's why it is advised to have a container of calcium available to the hens at all times. Good sources of calcium for hens include their own washed and crushed eggshell, or crushed oyster shell. I found that my girls had little interest in their cup of crushed shell during the day, but in the early evening, just before going to bed, they sought it out and consumed it ravenously.

Since chickens metabolize their food very quickly, my guess is that during the nighttime hours their internal egg factories begin the shell-making process in order to deliver a finished product the next morning. So going to bed with a full load of calcium on board is a necessity.

Although a shell made of calcium provides protection for the chick inside, it is the egg's exquisite shape, rather than its material, that makes it strong. The curve of an egg's shell works like an architectural arch or a dome. In structures like these, tremendous weight and pressure are distributed throughout the building material rather than being concentrated in one spot.

Pressure on the curved eggshell works in the same way, so that a monstrous hen like Lucy can place all her weight on the shell without disastrous results.

Of course, if Lucy were to pounce upon it, the egg might not be so lucky. A quick strike in one spot can easily break the egg, like when you whack it on the edge of a frying pan. In the same way, a block of concrete is very strong under steady pressure, but when I deliver my famous karate chop, I can crack it like an egg, no problem.

14

Pip

Inside the shell of a fertilized egg there is a tiny, virtually invisible zygote, which has the potential to become a chick.

THE EGG

SHELL

MEMBRANE

YOLK

ZYGOTE

The zygote is attached to that yellow yolk, which will provide all the nutrition for its twenty-one days of growth.

Outside the shell, the mother hen is responsible for warmth and humidity. Her body temperature, normally 102° or 103°F, drops to 100° and she loses the feathers in the center of her breast. This ensures that her body heat is transferred directly to the eggs. Her bare skin also transfers moisture to the egg, in order to maintain proper humidity within the shell.

A hen Lucy's size can easily manage a clutch of about a dozen eggs. When a hen does have a large clutch, she rearranges the eggs several times a day. The ones on the outside are moved to the inside and the inside eggs are rolled to the outside, to ensure uniform incubation. She also rolls each egg around to keep the baby from sticking to the inside of the shell. Since Lucy had only one egg to fuss over instead of a dozen, I wondered if our one chick was experiencing twelve times the attention.

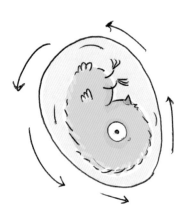

Lucy's wonderful chant, it turned out, was not a Zen mantra. It was a communication with her unborn chick. So Lucy had been building their bond from the start. I wondered if the chick had been listening to Hatsy and me and was bonding with us as well. And if I set up a stereo and piped in a bit of Mozart, would it boost our chick's IQ?

When I read that the baby starts peeping inside the shell a couple of days before hatching, I rushed out into the night to hear for myself. I reached under Lucy, found the warm egg, and put it to my ear.

Nothing.

The next night I read that if you tap on the shell, it will tap back. So out I went again, to sit in the dark under an umbrella in the rain with an egg to my ear.

I tapped.

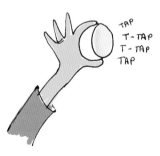

I heard it—a little tap-tapping inside.

I tapped again. Chick tapped again.

Morse code?

It gave me shivers. I put Lucy's egg back and pranced home smiling.

Bright and early the next morning:

Pip!

Chicken folk call that first little hole a pip, which is not to be confused with a zip.

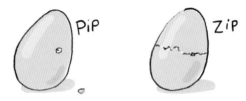

The pip is usually in the center of the shell, halfway between the ends, where the shell is thinnest. Pipping occurs when the chick's lungs begin to function. There is a small pocket of air inside the egg, but once that oxygen is used up, the chick begins to asphyxiate. Its neck muscles convulse and it flings the head forward, and . . . *pip!* An air hole, thank goodness, is created.

Once that breathing hole is established, the chick has an air supply. The chick then relaxes for a few hours or even for a whole day.

But I never did read the part about the chick relaxing. I read only as far as the pip. So after the pip, I assumed that Lucy's egg would very soon crack in two, and out would jump our fluffy new chick. I left Lucy with her pipped egg and returned twenty minutes later. Nothing. Twenty minutes after that: nothing. All morning long I pestered Lucy. She was calm, I grew more and more anxious, and the pip remained just a pip.

I would be teaching an art class in my studio that afternoon and my students were due to arrive soon, but I didn't want to miss the blessed event for which I'd waited twenty-one days. To stay out of Lucy's hair, I went inside and cleaned up the studio a bit and made preparations for class. Then I hurried back out to check on the egg one last time.

I carefully opened the nest box door,

and the egg flew out.
It fell two feet onto the hard bare ground and cracked.

Inside the egg, our baby screamed.

I picked it up and cupped my hands around the cracked shell. I broke into a sweat. Why had the egg been alone in the corner? Why had Lucy kicked it out of the nest? Was that normal? Had she been confused by the sounds inside the egg? Had my incessant annoying vigilance caused this?

I looked closely at the egg. The shell was smashed on one side. I saw a bit of blood and some wet feathers.

Then I heard laughter and looked up to see four art students marching toward me across the sunny lawn. I had to act fast.

Lucy was agitated, I was shaking, the egg was screaming. I placed it back in the middle of the nest, shut the door, wiped my brow, and rushed to my students.

"Did the egg hatch yet?"

"Nope, not yet—let's go do art!" I tried not to faint as I herded them away from the coop and into the house.

15

Egg Emergency

That was maybe the longest two-hour art class I ever taught. There was one small part of my consciousness that inspired and entertained four young artists, while all the rest of me squirmed at the thought of our little chick, maybe alive, maybe dead, maybe still screaming in its shell.

When class finally ended, I rushed the children out the door and into waiting cars. I waved and smiled as the last of them disappeared down the road, and then I turned toward Lucy's coop. I listened carefully as I walked toward it, but I heard no sounds.

This time I opened the nest box door oh so carefully and held my hand beneath it to catch any projectile.

Lucy was on the nest. She looked at me as she always did. Placid and content.

I slowly slipped my hand under her breast and helped her to stand. We both looked.

No egg.

No chick.

No sound.

I peered behind her and in the corners. I looked all around the coop. Nothing.

I slid my hand through the bedding and the hay but couldn't find a hint of anything—no shell, no feather, no blood. Then, while sifting around in the left corner, I felt the egg.

It was cold.

I lifted it out. Pine shavings were stuck to the side that was smashed. I delicately plucked them off. Looking closely, I could see a bit of brown gooey chick inside.

I cupped my hand around the cold shell and blew warm coffee breath on it. There was a tiny movement inside the shell. I shut the door on Lucy, who had sat down and resumed her brooding as if nothing had ever happened.

If this baby had any chance to live, I was going to have to take over for Lucy. Still breathing on the shell, I stood up and hurried toward the house.

But now what? Mother Nature had designed Lucy to be the perfect incubator. Okay. The two things the egg needed most were heat and humidity.

Sarah helped me set up our emergency room in the studio with a box, a towel, a squirt bottle of water, and my desk lamp.

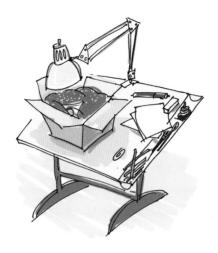

It was the best we could do in a pinch. With the squirt bottle, I dampened a paper towel and placed it over the egg to keep the desk lamp from cooking our baby. Sarah and I looked at each other. We peeked under the paper towel at the egg. Nothing was happening. She asked me if we might peel some of the shell away in order to help it hatch. I recalled reading somewhere that it is very important to allow the chick to hatch on its own, so we decided it would be best to wait and watch.

So we waited and watched. But the chick wasn't doing much of anything. Sarah eventually left the studio. I sat down. An hour passed and the chick seemed slightly more active. Another hour passed and I realized this was going to be a long process. I took a break and came back with a tall glass of chardonnay and a big bowl of potato salad.

Another hour passed.

Whenever I saw some movement, I lifted the paper towel. I glimpsed parts of a chick inside the membrane, but there were really no recognizable body parts. I couldn't even guess how this creature was folded so compactly inside its package.

At last I recognized the egg tooth: a pointy protrusion on the top of the beak that is used solely for hatching purposes and then falls off a day or two later.

Another eon elapsed.

Eventually a foot emerged . . . and another. They were enormous.

Sarah and Danny stopped in to check on us every hour or so. Marky came in and lay down on the floor beside me. Our chick struggled, napped, struggled some more, and finally broke out.

With one last kick, it was
free of the shell.

Then naptime again.

It was a redhead. The wet fluff looked like hair. The baby looked more like E.T. than a chick. I supposed it needed time to unfurl, like a new butterfly's wings when it emerges from the chrysalis.

All of its parts were present and accounted for. Wings, feet, even a comb. The disastrous fall from the coop that morning hadn't caused any evident damage.

Now what?

Well, I supposed that if the chick had hatched beneath the mother hen, it would get a nice buffing from her feathers and it would fluff right up. So I buffed with a tissue.

And I buffed some more.

There.

What a triumph. What a beautiful prize. I felt so happy for Lucy that she had been successful in her brooding.

Well, successful until this last day.

I sandwiched our new treasure carefully between my warm hands and took it downstairs to meet Danny and Sarah, but I allowed them only a moment of adoration before hugging the chick to my chest and scooting it out the door. This chick was

Lucy's, not ours. I hoped that Lucy would welcome it back. I carried it through the darkness to its home.

"Budup?" Lucy asked as I opened her door.

"Budup," she said again.

The instant the chick heard her voice it peeped frantically, hopped to its feet, and tottered off my hand onto the nest. Lucy's chant quickened. She lifted herself as high as she could and looked down as the chick dived into the warmth of her fluff. I closed her door and waited a moment, listening as hen and chick shifted around and settled in.

The budups continued, the peeping continued.

The peeps got louder.

The peeping sounded hysterical.

The budups seemed desperate.

I opened the door and lifted Lucy to see what was wrong.

Lucy's baby had cuddled into the crook of her knee. As Lucy folded her leg to sit, she was actually choking the wee thing. I lifted Lucy high enough for the gagging chick to free itself. It floundered away from her leg and crawled to a safer location beneath Lucy's hulking body.

Shortly after birth, a lesson learned: the struggle to survive never ends, even while nestled at the bosom of a loving mother.

I closed the door again.

The baby's muffled peeping quieted down and then ceased altogether, while Lucy's soft chant continued.

"Good night, Lucy," I whispered, and walked away.

16

Lucy Gets to Work

The next morning, the chick popped right out from under Lucy and tottered to the doorway to look at me.

I was overjoyed to find that Chickie had survived the night in the care of her less-than-graceful mother. She had fluffed up very nicely, too. I decided right there that the chick was a she because it's always best to have a positive outlook. Of course, only time would tell.

Chickie was active and bright eyed, and she turned her attention immediately from me to her mother. Lucy had managed to get to her feet without my help. She was poking around the nest, appearing to be busy and talking nonstop. She said all sorts of chicken words I had never heard before, and the chick listened attentively and responded promptly. Everywhere Lucy turned, her chick was right beside her, skittering to and fro to avoid being crushed by Lucy's gargantuan feet, all the while watching her mother's face and her movements. I placed a small cup of chick food in the nest box. Lucy pecked around in it, and when she found the one morsel that suited her, she changed her words to a higher-pitched *tut-tut-tut-tut.*

Chickie rushed to the dish and picked at the tidbit that Lucy held in her beak. That *tut-tut-tut* is called "tidbitting," and it is used to announce and offer food. This was new to me, but Chickie apparently understood from the start.

I had brought Lucy a special treat, too—a big slice of tomato. I placed it on the balcony beside the nest box and she came right out to enjoy it there. After eating a few bites herself, she announced the food with a "tut-tut-tut," and Chickie staggered eagerly to the threshold. Lucy continued her call, and Chickie lurched forward, overshot the tomato, and tumbled off the ledge into a pile of dry leaves below.

Lucy clucked hysterically to her missing chick. I opened the lower door and fished the chirping chick out of the leaves. Once Chickie was back by her side, Lucy calmed down and they finished the tomato together. Then Lucy returned to the nest box and sat herself down, and her baby dived in beneath her. Under Lucy, Chickie was safe, more or less, for the moment. But this coop was not a safe place to raise a chick. It was fraught with chickie perils. The water cups were drowning hazards, the ramp would be a challenge, and if the little nugget did any more catapulting off the ledge, she was bound to get hurt. Since I couldn't think of any way to make Lucy's coop chickproof, I was going to have to build something new.

I visited my ever-growing scrap pile in the garage and sifted through leftover lumber and fencing. Then I sketched some ideas and made a plan.

Simple.

HARDWARE
CLOTH

WIRE
CUTTERS

BANDAGES OR GLOVES
(your choice)

It was a cage made of (ouch) hardware cloth. It was about four feet square.

BROODY CAGE

Flaps on the bottom made it darn near impossible for a predator to squeeze in under the wire. A few bricks and rocks on top of those flaps ensured a very secure cage for hen and chick.

And since it was so portable, I would be able to drag the cage to fresh clean grass and to sunshine or shade anytime a change was necessary.

A storage bin inside the cage would serve as the broody house, a cozy place with no ramps, ledges, or sharp protrusions.

STORAGE BIN → BROODY HOUSE

Within a couple of hours, hen and chick were in their new home. Lucy, always appreciative of my attention to detail, looked it over and expressed her approval by planting herself inside the upturned bin and rearranging bits of fresh bedding around her. From where she sat she could keep an eye on her little one, and from the look of exhaustion on her face, I could tell she welcomed the chance to sit down. She summoned Chickie with a clear and unarguable *cluck*. Chickie obeyed by diving under Lucy for a nap.

I couldn't find the old waterer I had used when Lucy and the ladies were little, so I improvised.

ORDINARY ICE TRAY → DROWN-PROOF CHICK WATERER

And when a light drizzle began to fall, I improvised again with a sheet of plastic.

This setup would do just fine for a few weeks until Chickie grew and toughened up a bit. Then hen and chick could move back into the special-needs coop.

Chickie was such a delicate thing that at first I had a hard time leaving her outside in all sorts of weather and darkness and dangers,

with only Lucy to care for her.

But Lucy proved to be a confident and competent mother, and she gave her chick everything she needed.

Once everything was secure, I let Marky out to investigate the broody coop and its contents. He trotted past the new cage but did a double take when he spied the chick inside. At the first glimpse of Chickie, he regressed into the twitching drooling uncontrollable mess that he had been when the girls were tiny.

He just could not equate this: with this: ,

when this: looked so much like this:  .

So when Lucy and Chickie went out to explore the world,

Marky was kept at bay.

I knew Lil'White wouldn't take kindly to Lucy's chick either, so I blocked her exit while I let Hatsy out of the coop. Hatsy flapped joyfully across the yard to join Lucy in the garden. I followed right behind, eager to see her reaction when she first laid eyes on our beloved chick.

Hatsy pecked around beside Lucy for a moment before she spotted the chick behind a tuft of forget-me-nots. There was no love in her eyes when she took a running dive at Chickie.

Hatsy aimed to kill. Lucy lunged at Hatsy. Chickie squealed and plunged into Lucy's fluff and Hatsy circled around and around the two.

Boy, had I been mistaken. Apparently it was only the egg that Hatsy had so adored and not the chick within. I caught my little orange dynamo and hustled her back to the coop.

By the time I returned to Lucy and Chickie, they had recovered from Hatsy's attack and were back at work. Lucy was announcing all tasty findings with a *tut-tut-tut-tut,* and Chickie was scarfing them right down. Beetles, slugs, everything Lucy dug up, Chickie devoured. Everything except worms.

Chickie hated worms. Lucy loved worms. Lucy offered worms over and over, but her *tut-tut-tut*s just didn't work for worms. Chickie kept her beak shut tight until Lucy got the hint. Less than a day old, Chickie already had opinions.

I had been wondering if I would notice any differences in behavior between Chickie, who had a mother hen to raise her, and the four chicks I had raised in my living room. One distinction stood out right away: Chickie had quite the attention span. While my four living room babies had been a willy-nilly bunch, stepping all over each other and bouncing about, this chick of Lucy's had powerful focus—all eyes on mama, all the time.

Chickie responded to Lucy's every word and was ready at a moment's notice to accept bits of food from her, or to scuttle underneath her if Lucy sounded the alarm.

After a few days in the garden, Lucy led her little one on a greater tour of the yard.

I wondered where she found this new strength and stamina, when during those weeks of brooding she had been barely able to walk. I chalked it up to determination.

Lucy jetted across the lawn with her little one. She was a hen on a mission.

She taught Chickie the art of scratching and pecking. Chickie learned to stay close,

but not too close.

She taught Chickie to hunt for sorrel and dandelions, caterpillars and grubs, and to be respectful of bees, wasps, and ants.

Sometimes Chickie had to learn things the hard way.

As I followed hen and chick on their educational and nutritional forays, I observed another difference between Chickie and my original flock of living room chicks.

My indoor chicks ate chick starter feed. It was their dietary staple. This feed was specially formulated to provide them with proper levels of necessary nutrients. By contrast, Chickie barely touched the stuff. The starter feed was always available to Chickie, but Lucy filled her little one so full of grubs and beetles and sweet clover and grass that the chick had no room in her belly for scientifically engineered nutrition.

Lucy wore herself out with her rigorous home-schooling agenda. After a good half hour of lessons, she returned with the baby to the broody coop and pretty much passed out.

Chickie waited patiently while Lucy took her nap, and then they were off and running again.

Out on their forays, even while I was watching over the pair, Lucy was extremely vigilant. She knew that Chickie was the easiest of targets and a tasty treat for any predator.

Chickie needed no explanation of her mother's hawk warnings. At Lucy's first trill, Chickie dived for cover.

For additional protection, Chickie had some camouflage tricks of her own—she happened to do a great impression of a dead leaf.

Find Chickie in this picture:

I swear Chickie is in this picture too:

To provide her chick with a well-rounded education, Lucy traveled with her each day to more distant and exotic places, like the front yard. There, she taught her little one to climb stairs

in order to beg for treats at the front door.

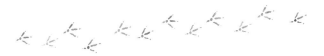

Back in their broody cage in the afternoon,

Lucy watched over Chickie from the comfort of her storage bin. Lucy was strict about bedtime. She required that they turn in way before dusk.

Some evenings, Chickie was just not into that.

Some days, Chickie was just not into walking, either.

Lucy was a good sport about it.

It took Chickie only a week or two to grow strong and smart enough to be able to live safely in Lucy's coop.

Chickie had great fun climbing up the ramp and jumping off the ledge.

Lucy kept on feeding grubs and bugs to Chickie as fast as she could, and Chickie grew very quickly. Her comb and tail feathers grew quickly also. Way too quickly.

My heart sank.

Chickie was a boy.

17

Lucy Returns

When I accepted those eggs from Patricia, I accepted the risk that we might hatch a boy. But at the time, I was caught up in Lucy's broody adventure and didn't want to think any unpleasant thoughts. Now I had a rooster again and I would have to find a home for him. But he was still a chick; he wouldn't be crowing for a while. In time we'd have to send him packing, but for now there was plenty of adventure still to enjoy. And besides, Lucy adored the little fella. Her enthusiasm in raising that chick completely offset the pain in her legs and the weakness that her illness had caused. I would keep Chickie for Lucy for as long as I could.

I expected him to be a little more independent by the time he was five or six weeks old, but he remained quite the mama's boy. Lucy was still feeding him, forgoing her own nutritional needs, and all that exhausting child-rearing  activity was taking an evident toll. As Chickie grew, Lucy shrank. She got skinnier and skinnier, and if that wasn't bad enough, she molted.

All birds molt. It's a natural occurrence. Feathers are astoundingly durable, but they do require replacement from time to time. All chickens have their own molting style. Some drop a feather here and there, and it's hardly noticeable. Other hens opt for the speed molt. One day they look voluptuous, and the next day the coop looks like a chicken exploded, feathers absolutely everywhere, and a nearly naked hen cowering in the corner.

Lil'White tends to fancy the explosion method.

But since she seems already to have twice the number of feathers she ought to have, she looks just as pretty the day after the molt as she did the day before.

This molt of Lucy's left her looking pretty scraggly. Her tail feathers were gone, and about half of all the other feathers as well. During a molt, the hen stops laying eggs and spends all her energy creating new feathers. Lucy hadn't resumed laying anyway, so that was fine. But as she was so worn-out from mothering, this molt was especially taxing on her health and energy.

Chickie stayed right by Lucy's side, keeping his mom company while she sat around looking miserable.

When Lucy found the energy to get up and move about, Chickie was right there, in order that his exhausted mother might cater to his every need.

Lucy was very tolerant of Chickie's separation issues. As her little man grew, the pair looked sillier and sillier: this gangly teenage boy still clinging to his mommy.

Weeks passed,

and Chickie did become more independent. And more gorgeous. He was so handsome that Lil'White took a real liking to him.

Hatsy still had no tolerance for the guy.

But the little whippersnapper could run faster than Hatsy could, so life was good. The whole flock was back free-ranging together again.

By now, Chickie had completely outgrown his name. I didn't want to give him a real chicken name because I felt that naming him would endear him to us just that much more and we'd have a more difficult time giving him up when the time came. But we had to call our rooster man *something,* so Roosterman he became.

Now that her boy was an elegant and independent rooster, Lucy found freedom to rest and recover. All of her feathers grew back, and her strength came back too.

In fact, she looked magnificent—prettier and stronger than she ever had looked.

And now that Roosterman was looking more like a chicken than a squeaky toy, Marky had no trouble recognizing him as one of the flock. So Marky was allowed to run free with the flock once again, and he dutifully resumed his chicken-watching post.

Roosterman learned right away that the white furry guy had a set of sharp fangs and that it was best to keep confrontations to a minimum.

On nice mornings, I propped open the doors to both coops, Marky stood watch over the flock, and I did my artwork at the kitchen table so that I could keep an eye on things through the window. Glancing out at the coop one day, I was stunned to see Lucy inside the big coop, ambling down the ramp and out onto the lawn. Lil'White stood right outside the door, and she sauntered amicably over to Lucy to graze beside her.

I blinked. Lil'White and Lucy were now friends? After all of Lil'White's hateful bludgeoning and vicious attacks? All of a sudden Lil'White had no problem at all with Lucy's walking in and out of the big coop, and Lucy had no fear of Lil'White. I felt as though I must have skipped a chapter. And on top of all that, Lucy managed to climb all the way up and back down that long ramp on her sore legs.

Moments later, Hatsy climbed out of a divot and joined her two friends. I watched the three hens pecking and scratching together in blissful harmony, and I was completely puzzled.

A short while after that, when I went out to collect eggs, I discovered one more surprise: Lucy had laid an egg in the big coop. Now I understood. This egg was Lucy's statement, her formal triumphant announcement that she had made a comeback.

That evening when the flock marched back home to roost, Lucy bypassed her own doorway and followed Hatsy and Lil'White into the big coop. She stumbled courageously up the ramp and joined Hatsy inside on the high roost. Lil'White accepted Lucy's roosting choice without demur and settled herself onto the lower roost, alone.

I was absolutely elated at all that I had witnessed. I cupped my hands against the henhouse window and peered inside. There she was. Lucy, my phoenix, had returned.

Meanwhile, Roosterman stood alone and confused between the two coops. He didn't know which way to turn or in which coop to sleep. I picked him up and tucked him in beside Lil'White on the lower roost. The clucking I heard as I locked things up assured me that there was peace in the henhouse.

18

Roosterman

The next morning I found piles of feathers all over the coop floor. They were rooster feathers, and this was no molt. This was Hatsy trying to pluck poor Roosterman naked. He could outrun her on the lawn, but locked up in the coop, the guy didn't stand a chance. I found him in the henhouse, cowering on a roost. Several of his impressive tail feathers were broken off and his ego was pretty badly damaged. I carried him outside past the glaring Hatsy and set him free on the grass.

Then Marky and I took a stroll through the woods in search of a special perch for our henpecked roo. We returned with a nice thick oak branch. I sanded it smooth and installed it high up in the coop, too high for Hatsy to jump. When the flock returned home that afternoon, our agile young man flapped up to the safety of his new roost, leaving Hatsy steaming and pacing below.

By now Roosterman was far bigger than his father had been, and he still had more growing to do. Lucy and Lil'White were impressed with his fine manly manners. When he discovered bugs and beetles in the lawn, he announced them with a *tut-tut-tut-tut,* tidbitting just like his mother had. He always proffered the best of his finds to the ladies. While they partook of his offerings, Roosterman circled them with a happy dance.

This is the way of the rooster. It's a courting behavior. Chivalrous and brave in the traditional rooster fashion, Roosterman stood tall and kept a keen eye out for danger, watching over the ladies while they grazed. He was as dedicated a protector as Marky was.

This proved to be a problem. Marky tolerated Roosterman just fine, but Roosterman had issues with Marky. I don't know whether our young rooster saw the dog as a rival or as a chicken predator, but he made it a point to torment poor Marky often and relentlessly.

He charged him.

He danced circles around him.

Marky displayed the patience of a saint.
But then Roosterman started playing dirty.

The first time our feathered fella launched an attack from the rear, Marky came running to me—confused, unnerved, humiliated. This was just plain not fair. What was Marky supposed to do? He was under strict orders to leave the chickens alone. But what about self-defense?

I didn't know how to solve this problem. So I told Marky he was a good boy and then locked Roosterman in Lucy's little coop for the rest of the day.

The next day Roosterman attacked Marky with the same strategy. This turned out to be a big mistake. Marky swirled around and caught his tormentor in midair and nailed him to the ground.

He would have sunk his fangs into Roosterman but he couldn't get a toothhold in the jumble of twitching feathers.

Lucky for Roosterman, I was nearby. I gently removed the dog from the great mound of rooster. Marky lowered his head and looked up at me sheepishly. He looked terribly sorry. I gave him a pat and a smile. "Good boy."

Roosterman, unscathed, hopped up and retreated. After a few shakes and preens, he resumed his proud strutting like nothing had happened. I fetched one of my handy-dandy chicken cages and threw it over him.

And from that moment, he was never a free man again.

I thought he would be devastated by his incarceration, but it didn't really cramp his style. From the cage, Roosterman still watched over his flock.

And from the cage he managed to summon his hens and keep them close. He tidbitted. He danced. I think the girls liked him even more now that he was locked up and couldn't inflict so much of his virile manliness on them. In the evenings, I allowed him to join the ladies in the coop, so he was a happy guy still.

Roosterman turned out to be a late crower. For the first few quiet months, I was in blissful denial about the inevitability of his mature voice. Who knew? Maybe this rooster would be mute. It could happen, right?

When Roosterman did eventually find his voice, it sounded more like a clearing of the throat than a cock-a-doodle-doo.

Certainly not unpleasant, and hardly louder than a whisper. Thank goodness Roosterman was going to be a soft-spoken fella, and I would be able to suspend my hunt for his new home. I was elated, especially because I hadn't yet begun the search. But Roosterman had only just begun to crow. It took a couple more weeks of practice for his voice to blossom into an ear-splitting off-key fingernails-on-a-chalkboard screech.

To stand near him when he belted out his song was utter agony. To those who lived a good block or two down the road, I guess it was tolerable, because when I went door-to-door apologizing for the noise, our neighbors were very kind and understanding. I promised them that Roosterman would be moving to a new home soon. The neighbors made it pretty clear that they looked forward to that peaceful day.

I began searching for a home for him—craigslist, Freecycle, chicken websites, Yahoo! groups, Facebook, flyers.

I expected the search to be tough. And it was.

As autumn neared, Roosterman started waking earlier and earlier, and he felt it necessary to inform the entire world. As I recall, his 4:00 a.m. announcements were the most agonizing.

Danny loved Roosterman as much as the rest of us, but he was losing patience, and losing precious sleep. Roosterman must go, sooner rather than later. Sarah once again required that this rooster find a safe and happy home, not a stewpot, and I assured her that he would. But I was stumped. In desperation I stuck a twenty-dollar bill in my pocket, stuck Roosterman in the backseat, and took him for a ride.

We visited all the different farms in our area. I knocked on doors. I wandered into barns. I chatted with farmers. One old geezer laughed heartily when he saw my little man. He pointed to three enormous roosters strutting about the barnyard and bet me that Roosterman wouldn't  last twenty-four hours with these guys. I hopped in the car and sped away. I offered the next farmer twenty bucks to take him, but she laughed too. Nobody wanted the handsome young fella in the backseat.

Back at home, I looked around the yard. There had to be a solution. Maybe there was something I had missed. And sure enough, there it was, right in front of me. Lucy's empty coop.

That was it! I dragged the vacant coop into the woods and raked an enormous pile of leaves over it.

I added a bowl of water and a bowl of food, and when evening fell I tucked Roosterman inside.

It worked like a charm. The next day was Saturday. The morning was quiet and peaceful and the whole neighborhood slept luxuriously late. Long after sunrise, as I approached the leaf-covered coop to let Roosterman out, I heard him crowing inside. A very muffled, barely audible song.

The soundproof coop was a success. My family was happy, the neighbors were thankful. I was now able to control the hour at which Roosterman would wake the entire world.

And Lucy got to keep her little man. I was pleased for Lucy. She loved him so. The caged-rooster setup worked quite well for her too. She sat beside him and adored him all she wanted, and when she had enough, she hobbled away to spend some girl time with her good friend Hatsy.

As Lucy ambled on painful toes toward the shade of the forsythia, Hatsy slowed her pace to walk beside her friend.

19

Cycles

Each season had its effect on the flock. Springtime inspired Lucy's bout of broodiness; molts often occur with the change of seasons, and there was a rhythm to Lucy's health as well. She spent sultry summer days panting in the shade, and her inactivity brought stiffness to her legs.

Sometimes I carried her inside to cool off in the kitchen.

Other times she tottered to a cool shady place in the garden.

In the coop, I placed extra containers of water, but I suspected that Lucy wasn't drinking as much as she ought to because it took her such effort to get to the water cups. So when I visited the girls I made sure to place cool water directly in front of Lucy. She dipped her beak right up to her eyeballs and guzzled and sputtered and wheezed and guzzled some more.

When I wasn't around to look after her on summer days, I went straight to the coop as soon as I got home, no matter how late it was. Even in the dark of night when all had retired to the roost, I would reach up and press the cup gently against Lucy's chest. She'd wake and greet me with a *bup*, and drink lustily.

Danny was also concerned for Lucy and all the girls during the steamy summer, and I knew I could count on him to ply the ladies with plenty of cold grapes and ice water when I wasn't around.

Hatsy's slender physique and sparseness of plumage served her well in the heat. She barely slowed down, spending much of her day zipping around snatching mosquitoes out of the heavy air.

Voluptuous Lil'White spent a good part of the summer panting and holding her wings out from her body in an effort to stay cool. She found dust baths in the shade to be quite refreshing, and she cooled her pretty toes by scratching around in moist soil. But unfortunately, her malevolent nature bubbled to the surface on one particularly hot, humid day.

Although Lil'White had amended her behavior toward Lucy and had graciously stepped back down to the bottom position in the pecking order, she harbored a hefty passive-aggressive streak in her dark little soul. On that extremely hot day, Lil'White sat herself down in the nest box and, just out of spite, decided to act broody.

Believe me, she was not broody.

She played this broody game only to annoy her flock mates. If the other ladies were impatiently waiting their turn to lay an egg, well, that just made Lil'White sit even longer. I had seen her broody game before and I knew how to call her bluff. All it took to get her off the nest was a handful of cracked corn or a few raisins, and in a flash she would be outside snatching up treats with no recollection of ever having been broody.

On that scorching day as I worked in the house, I looked out the window from time to time and noticed that Lil'White hadn't come out of the nest box—for hours. I went to check on her. When I opened the lid of the nest box, she looked up at me with bulging eyes and beak wide open, gasping. It was really hot in there.

I lifted her out and placed her on the grass. I offered her some water, but she could barely stand up, let alone drink. So I whisked her across the yard to Marky's brand-new supersized water bowl and plunked her feet in the cool water.

Marky wasn't too sure he liked that.

I put her back on the lawn, but she still wouldn't drink—she just sat there gasping. So I lifted her golden wings and patted water in her armpits. I was concerned for her, but I also sort of enjoyed this rare moment communing with my beautiful standoffish Orpington. She would never tolerate such attention when she was feeling well. After about twenty minutes, she was able to rise and take a drink. She was going to be okay.

From across the yard, Lucy had been watching Lil'White's drama. When I carried my golden diva to a shady area nearby, Lucy limped across the lawn and sat down beside her.

By now I had known Lucy long enough that such an act of kindness and concern should not have been surprising. But still I was surprised. You just don't meet a compassionate chicken every day.

Lil'White made no acknowledgment of Lucy's sweet gesture. She stood still just a moment or two longer, then turned and walked away. And that was okay, because she's Lil'White.

The next day I cut a new window in the henhouse for cross ventilation so Lil'White's future vindictive exploits would not do her in.

Autumn brought relief for the whole flock, and the cool days revitalized our Lucy. She never did muster the strength that she had found when Roosterman was a chick, but on these brisk days she managed to shuffle around and keep up with Hatsy pretty well.

But for Hatsy, autumn brought challenges. After those summer months of vivacious health, her strange illness came back. Her bad days were interspersed with stretches of good health that lasted just long enough for me to believe it wouldn't happen again. But every week or two, it crept up.

Each time the illness struck, Hatsy got quiet and sleepy. She stood with her head tucked and her eyes closed, unable or unwilling to move. The episodes lasted an hour or two. And each time, on the following morning, an enormous egg appeared in the nest box. These episodes always occurred late in the day, which indicated to me that the problem was related to her egg-laying cycle.

One evening just after the sun went down, I headed out to put Roosterman into his coop and to lock up the ladies. As I opened the kitchen door and stepped outside, I thought I could see two small figures in the center of the yard. When I got closer, I was surprised to find Lucy and Hatsy standing alone in the darkness.

Lucy looked up at me. *Bup.*

Hatsy was hunched and silent. This time her illness must have overcome her as she was heading back to the coop. While I was very concerned about Hatsy, I was also overwhelmed by Lucy's gesture of loyalty. As darkness closed in on the two hens, Lucy was faced with a choice. She could shuffle home with Lil'White to the safety of the coop, or she could remain with her friend. It had been Lucy's choice to stand beside Hatsy as darkness and danger crept in around them.

This time, Lucy's compassion brought me to tears.

I reached down and picked up our little Hatsy. I cradled her in my arms and walked her to the coop, very slowly, so that Lucy could keep up with us. Lil'White was already hunkered down in the henhouse on her own roost. I helped Lucy onto the higher one and placed Hatsy, eyes still shut, beside her dear friend.

20

Hatsy

In between her strange episodes, Hatsy was as gregarious and lively and social as she ever had been. She remained the flock's undisputed leader and an active participant in all backyard activities. And not just chicken activities.

Hatsy helped me with yard work,

she helped hunt for my car keys,

and she performed occasional secretarial duties.

Hatsy helped Danny clear the gutters,

and on weekend mornings the two of them enjoyed raisins and warm sunshine together.

But her spells continued, and they began to occur more frequently. When I discussed Hatsy's problem with Rosario, the vet who had helped me with Lucy's Marek's disease, she suggested that I bring Hatsy to her house during one of the episodes in order that she could witness it. So I did. I was apologetic for ringing her doorbell at dinnertime, but Rosario welcomed me, and I carried Hatsy into her kitchen. Rosario cleared a spot on the counter. We spread a dish towel there and placed Hatsy upon it. She stood quietly, head tucked, eyes closed.

Small children and dogs of various sizes chased each other around the kitchen, obviously ready for their dinner. Rosario gave a pot on the stove a quick stir and then turned to Hatsy for a little exam. Hatsy didn't respond at all to her pokes and prods. Rosario set her laptop beside my little orange hen and surfed the Internet for information or any hints at all.

She told me that Hatsy appeared to be in pain. I had suspected this but hated the thought. Some people don't believe that pain in animals is the same as pain in humans, but whatever kind of pain she was in, it was debilitating. Rosario agreed with me that Hatsy's reproductive system had blown a gasket. Something was amiss, and it was pretty serious. But Rosario was at a loss as to how to help her. We discussed the fact that Hatsy was a production hybrid, and that her rapid-fire egg laying probably had taken a toll on her health. I was saddened by the connection to our factory farm food system that Hatsy represented. Her breed was designed to burn the candle at both ends. She laid more eggs in one year than some other breeds lay in three. So, in a way, our society's demand for cheap eggs was ultimately the cause of Hatsy's problems. Rosario smiled sadly and gave Hatsy a gentle caress. I thanked her for allowing me to interrupt her family's dinner with yet another sick chicken. Then I tucked my hen into my arms and took her home.

Sarah was disappointed that we couldn't come up with a cure for Hatsy. She suggested that we set up a spot in the kitchen for her to spend the night, but I really felt that the best place for her was among her flock. Out in the coop, I set her gently on the roost beside Lucy, who bupped a quiet greeting.

The next day, I discovered something horrific in the nest box. It looked more like a giant wad of chewed gum than anything else. After Hatsy expelled that horror, she stopped laying eggs altogether. I felt relieved for her and hoped that this was the beginning of healthy times.

But the end of egg laying was not the end of Hatsy's troubles. In the dead of winter, on the coldest of days, Hatsy began to molt. So our little orange hen who hadn't an ounce of fat to keep her warm now had very few feathers either. Sarah and I went out to take a look at her and see what we might do to help her keep warm. Sarah suggested I crochet her a sweater. I thought a heat lamp might be sufficient. Lucy and Lil' White came over to greet us, but Hatsy stayed shivering in the corner.

While Sarah and I continued our discussion, Lucy turned and ambled toward her friend. She stumbled forward until their bodies met, and then she pressed even closer. Lucy turned to face us, all fluffed and smooshed against Hatsy.

And there she stayed, keeping her friend warm.

That afternoon I put a heat lamp out in the run. Hatsy appreciated the cozy red glow and huddled so close to the lamp that I wondered if I should come out and baste her.

One cold morning a few weeks later, I stood at the window and looked out at my flock. I could tell that something was different in the coop. I could see Hatsy standing hunched in the corner, her face toward the wall. Lil' White and Lucy stood beside her. Roosterman in his cage near the coop was just hanging around as well. He didn't crow. He didn't strut. Lucy and Lil' White weren't scratching or preening. They weren't doing anything.

Hatsy's flock knew.

Their vigil lasted about an hour. Then Lil'White and Lucy left Hatsy alone in the coop and quietly scratched together out in the yard. For the rest of the day, Hatsy stood alone, sometimes in the coop and sometimes upstairs in the henhouse. I kept an eye on Hatsy and the flock all day from a respectful distance. Only once that day did I choke back my sadness and visit her. I offered her a bit of food, but she didn't eat. I offered water, and she opened her eyes and took a sip.

The next morning I found Hatsy sleeping on the floor of the henhouse. I petted her, but she didn't respond. She was blocking the doorway, and I was concerned that if the other girls needed to get to the nest box they might jostle or step on her. So I picked up my little orange hen and brought her back to the house. Danny met me at the door with a cardboard carton. After we added some pine bedding, I placed Hatsy gently inside. Danny sat down beside the box and reached in to stroke her feathers.

I started across the kitchen thinking maybe I'd put on a pot of coffee, but I heard a muffled flutter of wings and turned to look at the box. The fluttering slowed and then there was silence. Danny looked up at me. A puff of pine shavings snowed quietly down around him.

Our Hatsy was gone.

I wrapped her beautiful body in a clean linen dish towel and put her out on the screened porch. Then I got my garden spade from the shed and used it to slowly poke my way around the perimeter of the yard, searching for a thawed piece of earth in which to dig. When I found some workable soil under a pine tree, I fetched Hatsy's shrouded body and placed it on the ground nearby. Lucy and Lil'White came to join me.

While I dug the hole, Lil'White and Lucy eagerly assisted, lunging for each worm and grub as it was uncovered. The two girls showed no sentiment and didn't stop to look at Hatsy's orange feathers poking out from under the linen shroud beside them. Lucy and Lil'White had paid their respects the day before.

As soon as they enjoyed the last worm at the graveside, they moved on. I placed our dear Hatsy's body in the ground and covered it with soil. Then I leaned on my spade and fell completely apart. I cried and cried, more than I would ever have expected to cry over a chicken.

I cried for the end of a beautiful friendship between two soulful creatures. I cried for Lucy's loss.

The next morning, I watched from the kitchen window as my two remaining ladies busied themselves in the yard. Only two hens. The sight seemed unbearably wrong to me. Two hens just did not make a flock. The girls needed a third. And I needed to stop crying.

I got into the car and drove off in search of a hen.

21

Pigeon the Chicken

The nearest farm was only ten minutes away, so that was my first stop. I drove up the lane past several tidy pastures and stopped beside a dazzling white barn. Dozens of chickens of all colors and sizes drifted across the barnyard while two swarthy roosters kept watch. The farmer was fiddling with a big piece of machinery, and when I stepped out of the car he wiped his hands on his pants and came over to greet me. I clutched a tear-soaked tissue and took a deep breath, hoping my voice would come out steady. I smiled and introduced myself as a neighbor. I told him that I had just lost a chicken and was hoping to acquire a young laying hen to replace her, and asked if he might have a hen that he would be willing to sell. The farmer pulled off his cap and scratched his head and looked around at his large flock.

"No."

He was a man of few words. I broke the silence by asking him if he knew where I might be able to buy a young hen. He suggested I try a farm a few roads up, around three or four curves and past a field of sheep on the left.

So there I went.

Sure enough, just past the sheep I spied a fat brown hen scooting across a deeply rutted driveway. I pulled over and got out of the car. In the open doorway of a barn-type garage sat a skinny old man on a bale of hay. Beside him in the garage were half a dozen roosters in large stacked cages crowing at the tops of their lungs. A cacophony of chickens squawked and clucked somewhere out of sight.

Trying to appear both neighborlike and farmerlike, I tucked both hands in my pockets and sauntered casually toward the old guy. I smiled and complimented him on the pretty little hen in the driveway.

He adjusted his dirty glasses and peered at the hen.

"She's the neighbor's."

Another farmer of few words.

I introduced myself and explained that I was seeking to purchase a hen and that he had been recommended by a farmer up the road. He just squinted at me, so I proceeded to tell him about Hatsy's death and my quest for a new flock member.

He looked down. Said maybe there was one hen he could part with.

Slowly he raised himself up and turned around. I followed him into the darkness. We squeezed past an ancient dusty tractor and some other old machinery, past a row of chicken-filled cages toward a narrow staircase on the back wall. I stopped beside a hutch at the base of the stairs while the farmer climbed to the second floor. He opened a gate at the top of the stairs and disappeared. His footsteps met with a mess of flapping and squawking and carrying on, and fine dust powdered down from the rafters above my head. Then the gate swung open again, and out of a great cloud of grime emerged an upside-down chicken, followed by the old guy clutching its ankles.

The man carried the bird back to his hay bale in the sun. He flipped her right side up on his lap and attempted to make her more attractive by smoothing her disheveled and broken feathers.

"She's molting," he explained.

"Her comb's kind of mangled and bloody." I winced.

"Rooster did it."

"Her toenails are really long," I pointed out.

He pulled a pair of toenail clippers out of his pocket, clipped her nails, then put her down on the dirt.

The hen just stood there.

He told me that she was a Barred Rock, and that she was very friendly. He said that a boy up the road had raised her as a pet, but for some reason he wasn't able to keep her, so she ended up here.

My guess is that the old man had tossed her into a crowded cage with an established group of hens, and they had pecked and bullied her into her present condition.

I knew that I shouldn't bring such a sickly specimen home, but I had a feeling that if I didn't take her she'd be dead soon.

"How much?" I asked.

"Oh, how about five."

While I fished some money out of the car, the old man dug up a small carton and cut a breathing hole in it. He placed the bird inside and tied the package neatly with a length of orange twine. He handed me the box and I gave him the five dollars. I returned to my car with the little package, feeling rather foolish. The box was as light as a feather, and I found it hard to believe that there was even a chicken inside.

I sat it down beside me on the passenger seat and peeked into the hole.

A little eyeball looked back at me.

Back at home, I sat down in the garage with my parcel and opened it up to have a closer look. The bird didn't flutter or flinch as I reached in and lifted her from the box. I placed her on the floor and she took a few shuffling steps. Skinny and bloody, broken feathers, toes curled backward. She walked like a pigeon. She was a fixer-upper.

I offered her a small cup of water and she hesitated with every sip, lifting her head and looking around as if expecting to be murdered at any moment.

It was clear that she had been at the very bottom of the pecking order in that old attic. And she smelled really, really *bad*. I left her in the garage under an upturned laundry basket until I could figure out what to do next.

I decided to e-mail my friend Terry Golson, knowing she'd have some good advice. Terry is a chicken guru of sorts, whose popular Hencam.com blog I had been following for quite some time. She and I lived within close enough range of each other that we were able to get together for an occasional cup of coffee, and we had become friends in chicken fanaticism. I hesitate to bother Terry for advice because she has thousands of followers who probably also bother her for advice. But as usual, Terry responded to my e-mail right away. She told me to give the hen a good bath and to put her in quarantine.

A bath?

Yep, and a blow-dry. Terry told me how.

Pigeon actually seemed to enjoy her spa treatment, although it completely wore her out.

Afterward, she still looked really bad, but she didn't smell so much.

I wheeled the chicken tractor around to the front yard and set it up as Pigeon's home for her month of rehab and quarantine. During this period I would watch her for any coughing or mites or signs of illness that could endanger my healthy girls.

I did have concerns about her condition. Her breathing seemed raspy, she was terribly skinny, and she continued to shuffle around with a hesitating step, in a squatting position.

Danny and Sarah came out to meet the new arrival.

"Is she going to live?" asked Sarah.

Danny offered Pigeon a piece of clover through the fencing, and she pecked at it.

"I don't know." I really didn't know whether she'd survive. The best I could do was feed her and keep her comfortable.

I showered the little hen with all sorts of foods and treats. She didn't know what to make of the apple slice. She peered into the feed cup and chose a small morsel, then looked all around to see if somebody was going to take it from her.

After a couple of days I opened the tractor door and invited her to explore the yard. She wasn't too keen on leaving the safety of the tractor, but I was eventually able to lure her out with food and kind words.

The big sky overhead frightened her,

so she hunkered down beneath a shrub.

I tried to engage the hesitant hen, and after several days of good meals and gentle treatment she began to show interest in her new environment.

I delighted in introducing her to the finer things in life.

Pasta bewildered her.

She couldn't figure out what to do with a raisin either.

One morning Pigeon met her first slice of tomato. After some hesitation, she pecked it. And if a chicken can smile, she did. She glanced nervously around, fully expecting someone to come take this precious treasure—but no one did.

Pigeon had found heaven.

When the weather warmed up, I enticed Pigeon out of her tractor by digging in the front garden with my trowel.

No chicken can resist freshly turned soil.

She stood beside me and watched, but her scratching-and-pecking skills were rusty.

I found a plump grub and tossed it in front of her. She didn't see it. I picked it up and held it in my palm. She examined it closely with one eye, then pecked to the left of it. She looked again and pecked again and again, and finally hit her target. I wondered if there was a problem with her vision, and if this had contributed to her rank at the bottom of the pecking order in the attic.

Pigeon waited eagerly for me to unearth another treasure, but she didn't scratch in the soil. Rather than watch the trowel, Pigeon watched my face, just as little Roosterman had watched his mother's face.

What an endearing mystery this little hen was.

Marky was curious about the newcomer, and was eager to come over for a polite sniff. But I asked him to keep his distance while I worked with Pigeon in the front yard, so he did.

Pigeon acquired a measure of trust in the big white guy with the teeth, but she remained wary. She uttered "Oh-oh" to warn me of his proximity, just like the other hens had always done.

As the weeks passed, Pigeon's fears faded away.

Scabs faded, too, and new feathers emerged.

My sadness over Hatsy began to fade as well. Pigeon was a comical companion. Her rehabilitation and healing were inspirational.

22

Pigeon Joins the Flock

When I wasn't in the front yard with Pigeon, I was in the backyard observing how Lucy, Lil'White, and Roosterman were adjusting to life without their leader.

My heart still ached, and I had expected that Lucy and I would share our sadness together. But among the chickens I found myself alone in grief.

Lucy, Lil'White, and Roosterman adjusted just fine. They carried on. There was no rearranging of pecking order between the hens. There was no expression of sadness. There was no tear in Lucy's eye.

Lucy wasn't the type to drop facedown on the lawn and cry her eyes out over the loss of her friend. Yes, Hatsy was gone. But there were still grubs to be eaten and clover to pick. And the fox and the hawk were still patrolling in the shadows. So, blessed with the brain of a chicken, Lucy let go of her friend Hatsy and moved on.

Meanwhile, in the front yard, Pigeon bubbled with enthusiasm and curiosity and questions. She adored Danny and Sarah and the FedEx guy and any friend or stranger who happened by.

She sought adventure.

Mystery and excitement lurked just about everywhere.

The garden hose was always a thrill.

One day she met a rabbit who tolerated her enthusiasm for several minutes.

When there was nothing else to do, Pigeon ran.

Every day for Pigeon was the best day ever.

In just a couple of weeks, she filled out and feathered out and became a pretty young lady. She hadn't yet laid any eggs, but I supposed she would work on those details when she was ready.

Still I had my doubts about integrating her into the flock. Would she be able to stand up to Lucy and Lil'White? Would my evil Orpington light into Pigeon and cause her to regress into the cowering pathetic bird I had discovered only a few weeks ago?

One afternoon as Pigeon helped me with a building project in the driveway, she looked up and froze.

Beyond the retaining wall, she had spied a mysterious golden beauty. The figment floated briefly among the irises and then disappeared behind the house.

Pigeon kept watching, but the lovely creature had vanished.

This was just too exciting for the both of us. Pigeon was in her final days of quarantine, and I couldn't wait any longer to get on with our adventure.

The very next morning, I marched Pigeon to the backyard and she followed me like a puppy. As Pigeon and I rounded the corner, Lil'White and Lucy stood tall inside the coop and stared. Roosterman, in his cage, dropped one wing and did a little dance.

Pigeon puffed out her chest, lifted her tail, and instantly looked like a real chicken. She scurried on over to his cage to meet him.

I placed an empty cage over Pigeon and then opened the coop door for the girls. Lucy stepped out and eyed the little Barred Rock from a distance, while Lil'White completely ignored her.

Pigeon attempted a bit of posturing for the two strangers. She fluffed out her neck feathers and stepped from side to side, but her gestures had no effect on the girls. So she turned and flirted with Roosterman, and he entertained her tirelessly with his strutting and dancing and torturous song.

The next day I sprinkled chicken scratch on the lawn, put all three girls out there together, and stood by to rescue little Pigeon if necessary. Pigeon puffed herself up and prepared for battle.

She approached Lil'White and growled.

Lil'White turned around and made a beeline for the forsythia.

Pigeon turned to Lucy and growled.

Lucy sat down. Pigeon pecked her on the head.

"oooWip?" said Lucy.

But Pigeon wouldn't take *oooWip?* for an answer. She kept on pecking. Lucy's legs were hurting her that day so she couldn't get away if she tried. Pigeon's pecking was relentless.

I just couldn't stand by and watch, so I gently inserted my garden rake between the two. Lucy stood up and Pigeon darted around the rake and squeezed in to get another peck at her. Wedged in there between Lucy and the rake, Pigeon took a good close look at Lucy's plumage. She plucked a speck of dirt off Lucy's back and then began to preen her.

I removed the rake. Lucy remained politely still, and Pigeon stayed at Lucy's side, myopically examining Lucy's feathers. When she was satisfied with whatever she had discovered, Pigeon gave her own scruffy plumage a triumphant shake.

And that was it. Lucy and Lil'White pretty much handed the throne to Pigeon. There was no battle, no interview, no discussion, no nothing.

Of all the outcomes I had envisioned, this one wasn't even on the list.

Pigeon stood tall and surveyed her new realm. There was a lot of exploring to do. First, she visited Roosterman, who danced for his new queen. Then she checked out the coop. She spied the enormous feed bin and rushed right over. The prospect of unlimited food made her giddy, and all table manners were abandoned as she dived in.

Lil'White, now curious about this newcomer, followed Pigeon into the coop and watched the new flock leader shovel food all over the floor.

Finally stuffed, and a bit full of herself, too, Pigeon turned to confront Lil'White for the second time. Lil'White submissively dipped her head below the level of Pigeon's and held her pose until Pigeon had passed by.

I sat beside Lucy in the warm sunshine, and Pigeon flapped and zigzagged across the lawn to meet us. She looked up at me.

"Bwip?" Then she gave Lucy a peck on the head.

Lucy bupped. She rose stiffly to her feet and plucked a blade of grass.

Pigeon began to graze beside her.

Marky pranced over and asked me to throw his squeaky toy,

but I told him "no."

I didn't want to miss a moment of this.

23

A Flock Once Again

As dusk fell, Lucy and Lil'White climbed the ladder to the henhouse and took their usual places on the roosts. Pigeon wandered around out in the coop down below and found a twig on the ground to sit on. I picked her up and placed her in the henhouse with her flock.

The next morning, I gave her a little lesson on climbing the ladder. It did no good. Every night Pigeon found herself another stick on the ground and sat on it until I went out and scooped her up and tucked her in. After a couple of weeks, however, she figured out the ramp on her own. Placing one foot haltingly in front of the other, her head lowered, she scrutinized each rung. Her poor vision was certainly part of the problem, but her bent toes didn't help things either.

A chicken's everyday scratching in the soil will normally wear her toenails down to a healthy length. But in Pigeon's previous home in the attic, she had had no such opportunity. Her overgrown nails had caused the toes to curl back like petals of a fleur-de-lis, so hers were not the most functional of feet.

For the first few days of Pigeon's reign, I gardened and watered and puttered around, and watched the flock's activities and interactions out of the corner of my eye.

Lil'White drifted in and out of the picture in her aloof Lil'White style. She and Pigeon didn't really have much to say to each other. But Lucy was different. Or . . . the same.

And Pigeon was smitten with Lucy.

Pigeon seemed to have decided right from the start that she was going to be Lucy's very best friend.

She followed the enormous tottering
Barred Rock wherever she wandered.

Everywhere.

Always.

They visited their favorite fella together.

When Lucy needed to rest, Pigeon milled about nearby.

While Pigeon milled nearby, Lucy watched for hawks.

Lucy didn't seem too thrilled to have this little shadow always at her heels. I think she would have liked Pigeon to give her a bit more personal space. But she tolerated the enthusiastic little hen and began to enjoy the attention.

In time Pigeon did acquire some confidence in her own chickenhood and began to give Lucy some breathing room.

And after a few months, our little fixer-upper evolved into a striking beauty.

She started laying eggs, too. And like the rest of the gals, Pigeon put her own personal spin on her product. Pigeon's eggs looked just the way I'd have expected them to look.

Unique. Every one of them.

People ask me how I can tell Pigeon and Lucy apart. I guess that if I had fifty Barred Rocks, I wouldn't be able to tell them apart, let alone know them as individuals. But with only two, the differences are like night and day.

Pigeon is compact and cute, with a Rubenesque figure.

Lucy, on the other hand, is a gargantuan and elegant bird.

Their conversations:

Pigeon ends every word with a question mark, while Lucy moans and laments.

Chickens recognize each other by their faces and the shape of their combs, and I can see that Lil'White has no problem telling the two Barred Rocks apart.

The girls recognize their favorite humans by faces and shapes, too.

The one time I wore a big floppy hat out to the coop, my ladies panicked. I removed the hat from my head, and when they saw my familiar frizzy mess of hair, they calmed right down.

When Danny comes out to visit, the girls recognize him instantly, and the three of them come thundering across the lawn to greet the Raisin Man, also known as the Stale Bagel Man and the Tomato Man.

I've done some rearranging in the chicken yard. There are some nice logs scattered around for Lucy to perch on, and I've transplanted fresh sod and weeds in the areas that had been scratched to bare soil. This year the ladies have their own magnificent tomato plant, which they grew themselves from a seed they must have missed. And with the soil so well fertilized by the flock, the plant is lush and abundant with fruit. I wrapped a fence around it for its own safety, and as the tomatoes ripen I feed them directly to the ladies. After all, they're the ones who grew it.

I've added to the doghouse setup as well. Lucy has taken to sunning herself in there on cool mornings, so I built a platform to raise it up and improve her view. The doghouse now sits

comfortably above winter snowdrifts and spring mud, and the structure is a favorite hangout for the whole flock.

Pigeon often squeezes in for a nap beside her best friend, and Lucy no longer appears to be annoyed by Pigeon's need to be close. In fact, I've noticed Lucy doing some of the squeezing herself.

It's not an easy job to be the very best friend of a behemoth like Lucy.

At the end of the day as the sun sets, Pigeon is always the first to go to roost. Once she's settled in, she calls to her flock and they file right up the ramp to join her.

On a recent evening, Lucy's legs were giving her a hard time and she didn't have the strength to hop onto the roost inside the henhouse. Not wanting to sleep on the floor, Lucy hobbled back out into the coop to sit on an outdoor perch. After a while, Pigeon got down off her roost and went looking for Lucy. Out in the darkness, Pigeon carefully stepped onto the perch and sat down beside her friend.

That night when I went out to put Roosterman to bed and to lock everything up, I found these two beautiful ladies roosting outside together.

Two round shadows.

Two Barred Rocks.

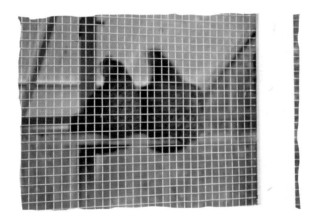

Two compassionate hearts.

I remember, with a lump in my throat, the special bond between Lucy and Hatsy. I still can't comprehend it—can't put it into words. It's something beautiful that simply *was*.

But perhaps I do understand this new connection between Lucy and Pigeon. I imagine that their personal hardships have fostered their sensitivity and compassion. On the other hand, maybe they both just happened to arrive on this earth equipped with beautiful caring souls.

I must remind myself that it's probably sweeter to sit back and observe their evolving friendship as something that just *is*.

Lucy taught me that.

And what's with Lil'White? Is she really evil, bitter, hopelessly self-absorbed, and borderline psychotic, or is she only misunderstood?

Again, a lesson from Lucy: I'll let the questions go, and just enjoy what is. The chickens have filled my backyard with more hilarity and drama and life than I ever could have imagined, and the adventures continue. So I'll pull up a stool and sit in the sun with my girls.

Because with a flock like mine, there's little reason ever to be indoors.

To my husband, Danny, for his dedication, faith, support and love.

To my father, for providing me with a lifetime of inspiration.

To my mother, supplier of fresh crayons and enthusiasm.

To Beth Towne, whose backyard asylum thrills me and who is always there for Marky and the ladies.

To Patricia Giglio, who inspires me.

To Terry Golson, Chicken Guru.

To Meg Cherchia, who helps me stay on the path.

To Laurie Abkemeier, my immensely talented navigator and cheerleader.

To Leslie Meredith, for envisioning this book and making it happen.

About the Author

Lauren Scheuer (LaurenScheuer.com) is a graduate of UCLA with a degree in Fine Art. She has illustrated more than a dozen children's books, cookbooks, and activity books and has taught illustration classes at Massachusetts College of Art and Design. Lauren's licensed artwork has appeared on puzzles, games, paper goods, greeting cards, and more. She lives in rural Massachusetts with her husband, their teenage daughter, Marky the terrier, and her flock of remarkable individuals.